Dedicated to my 1986 soul mates who stimulated my introspection with their sense of belonging and humour.

Copyright © 2021 DR RAJAN PULLANGHAD

All rights reserved

The characters and events portrayed in this book are fictitious. Any similarity to real persons, living or dead, is coincidental and not intended by the author.

No part of this book may be reproduced, or stored in a retrieval system, or transmitted in any form or by any means, electronic, mechanical, photocopying, recording, or otherwise, without express written permission of the publisher.

ISBN: 9798664626841

Printed in the United States of America

# CONTENTS

Copyright
FROM AUTHOR'S DESK
SUN: THE COSMIC POWER     1
WHY SOLAR ENERGY?     10
SOLAR RADIATIONS     15
SOLAR ENERGY RESOURCE ASSESSMENT     23
PhOTO VOLTAIC POWER     36
CRYSTALLINE TECHNOLOGY     47
COMPONENTS OF SOLAR PLANTS     55
SOLAR INVERTOR     62
ANTI ISLANDING PROTECTION     80
ROOF TOP SOLAR SYSTEM     91
ROOF TOP SOLAR INSTALLATION     97
Afterword     104
About The Author     106

# FROM AUTHOR'S DESK

Man has finally realised his long-cherished dream. He had been looking forward to this day when he could bring the supreme power ruling this earth under his control. He now no longer looked at the Sun in awe or wonder. He has now tremendous knowledge to make use of its powerful rays coming from space capable of producing heat and energy.

The present and future anticipated energy demand of world is only a small speck against the total incident energy from sun. This enormously excess energy can meet the total energy demand of generations to come. Highly diffused and pollution free source will trigger the world energy sector to heights if proper technologies are developed to harness it.

In stark contrast to the finite conventional, polluting energy derived out of fossil fuels, solar energy is a promising green energy source on which coming generation will certainly depend as it is eco- friendly, benign and omnipresent. The amount of solar energy falling on earth is enormous, as it is 200,000 times the whole electricity generating plant capacity volume of the world.

The decision to convert solar energy to thermal energy or electrical energy, decides the method to be adopted for. The cost involved in collection, conversion, and storage of solar energy limits its exploitation in many places though energy from the sun is freely available.

It will be an excellent alternative source of electricity which is a relief in the backdrop of fast reducing stores of our fossil fuels.

Currently 30% of energy demand is fulfilled by solar energy. Government funded and sponsored solar electricity generation is increasing drastically in developing countries.

I have tried with utmost sincerity to include in this book the components and operational details of roof top solar plants.

I have also tried to bring to my readers, an overview of establishing solar plant on our own rooftops or in a common area for communities comprising of many families. One of the fast-emerging renewable energy today is Solar and its advantage over other energy sources is obvious worldwide.

DR RAJAN PULLANGHAD

# SOLAR POWER

## ENERGY AT YOUR DOOR STEP

DR RAJAN PULLANGHAD

# SUN: THE COSMIC POWER

The Sun has always been an object of wonder and speculation to mankind. The Sun in our minds is a bright golden ball of fire with its rays spreading all over this world. There is no other force in the world that has influenced man as much as the Sun. It has been a symbol of power, growth, health, passion and the cycle of life. The Sun was once upon a time worshiped as a God with miraculous power. People believed that Sun was the Lord of the Earth. Even today, few communities in our society are still worshipping the Sun as God.

Breaking the boundaries of myth, we began to study the solar system. The exact time of origin of Sun is in dispute among scientists and it is calculated that four and a half billion years have passed since the origin of Sun. Although there is no clear record of the formation of Sun, we have the Sun's rotation time and its energy production to study about its physical and chemical properties.

Scientific explanations and propositions are the only basis to assume the origin of the Sun and the Solar system. Sun is calculated to be positioned at a remoteness of 147.58 million km from the Earth. Sun is believed to be in in plasma state with high temperature.

The size of Sun is about 109 times the diameter of Earth. It indicates a size more than three hundred and thirty thousand times more than our planet. In terms of size, the Sun embraces 99.86% of the total mass of the solar system. Concerning the structure of the Sun, two-third of its space, about 73 percent, is occupied with hydrogen. It also has 25 percent of helium gas and about 2 percent of carbon, oxygen and iron. About half of the total incoming energy from the Sun is absorbed by the earth. The remaining is reflected back into space.

**Mythology around the Sun**

There are countless references in books on ancient history regarding the Sun deity. Once Sun was worshipped as a God by the ancient civilization and was believed to illuminate the planet and nourish life on earth. For the prehistoric Greeks, Romans, Indians and Egyptians, Sun was an object of admiration and surprise, possessing curative powers.

In Mesopotamian era, Sun was considered as a sign of royalty. In Latin American culture, the Sun was well-thought-out as the soul of fire. Sun was adored by the Japanese as amethyst, and the Egyptians worshipped the Sun as Ra. In Roman culture the first day of the week was considered to be the day of the Sun, and hence baptized as Sunday. Figure shows the picture of the Sun chariot from Trenholm Denmark.

In ancient times, the Romans were habituated to receiving sunlight in morning hours. In Jewish culture and Christian beliefs, the Sun was considered a source of life. Christian churches of all ages were built in such a way as to receive morning sunrays and its bright light. In Australia, too, some tribes choose their king from the Sun dynasty.

In India, the ancients Hindus worshipped the Sun as the God Aditya. In Indian culture, the Sun was designated as the face of the Sun God. Sun was seen in Indian literature as a symbol of magnificence and supremacy.

The Sun is worshiped at the Konark temple in Orissa, one of the world heritages sites in India. The Gayathri mantra written in Indian scriptures was meant to please the Sun God. In the

Indian epic named Mahabharata, the birth of Karna, the eldest son of queen Kunti was believed to be with the blessingsof the Sun God, and was hence called Suryaputhra.

In history, we hear the stories of the Sun traveling in a chariot with seven horses as king. The seven horses represent the seven colours of Sunlight. It can be inferred from historical manuscripts that from the start of anthropological life on earth, many examples which disclose the dependency of man on solar energy.

**Sun salutation**

In ancient times, the Sun was worshiped as the symbol of the wellbeing and energy. Many ancient forms of exercises were associated with the Sun. Surya namaskar is one of the most important form among them. Surya namaskar is usually performed in the morning hours, facing the Sun.

Surya namaskar is one of the most important form among them. Surya namaskar is usually performed in the morning hours, facing the Sun.

The practice of shaping life like the rising Sun was popular at that time. The exercise method of Sun salutation has been accepted by the World Health Organization recently. Each posture in the Sun prayer is designed to improve the functions and blood circulation in all the organs of our body.

## Sun light: The effect on the human body

The Solar radiation includes visible light, Ultraviolet light, Infrared waves, radio waves, X-rays and Gamma rays. Sun light has both beneficial and deleterious effect on man. The most essential role of sunlight is its ability to enhance the body's vitamin D reserves. The commonest cause of vitamin D deficiency is due to lack of sunlight exposure.

The human body produces vitamin D through a series of chemical reactions with the help of sunlight. Vitamin D is essential for the growth of bones and various other processes.

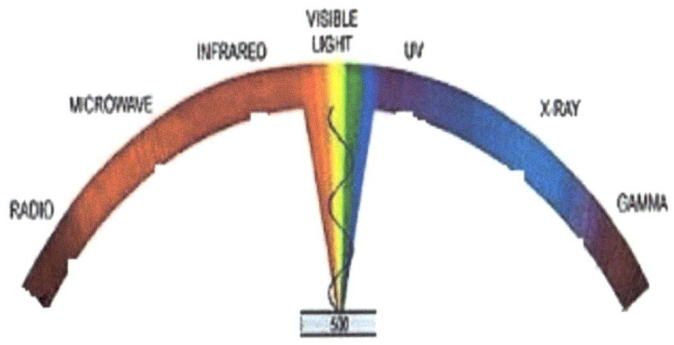

Vitamin D deficiency is very common among people of Australia, New Zealand and among Scandinavian countries as exposure to sunlight is very minimal at these countries. Thousands of genes governing each tissue in the human body are controlled by vitamin $D_3$ and it is the agent of calcium and phosphorous metabolism that helps the functioning of neuromuscular and immune system. Vitamin D is synthesised in human body by exposure to UVB radiations.

The amount of vitamin D production is dependent upon the type of dress worn, the amount of excess body fat, use of barriers like suntan lotion that block the radiations infiltrate the skin. The type of skin also decides the infiltration rate and the amount of vitamin production. So, deficiency is more common in fair skinned persons than in dark skinned ones. The vitamin D deficiency induced malfunctioning of bones causes growth retardation, skeletal deformities, rickets and bowed legs.

The practice of sunbath for skin tan is a fashion and a medical cure against vitamin D induced diseases. It is still a practice in our early new born care, to keep the new born exposed to sunlight for a few hours as a treatment against neonatal jaundice. This enhances the absorption of Bilirubin from the skin, thus leading to early cure. UV rays are used in the New born ICU for Phototherapy for the above-mentioned condition. The other beneficial effects are

- It is believed to increase the release of a chemical substance in our body called Seratonin, which is a mood elevator. It helps us to stay calm and focussed.
- Believed to be helpful in treating Seasonal affective disorder also called Seasonal depression.

The deleterious effects include

- Heat exhaustion and Heat Stroke

- Sunburn is caused by excessive exposure to Sun light and leads to skin blisters as seen in the picture. It can also cause headache, fever with chills, dizziness, nausea and dehydration.

- The collagen fibres are affected due to excessive exposure and destruction of vitamin A in skin and also cause premature aging of the skin.

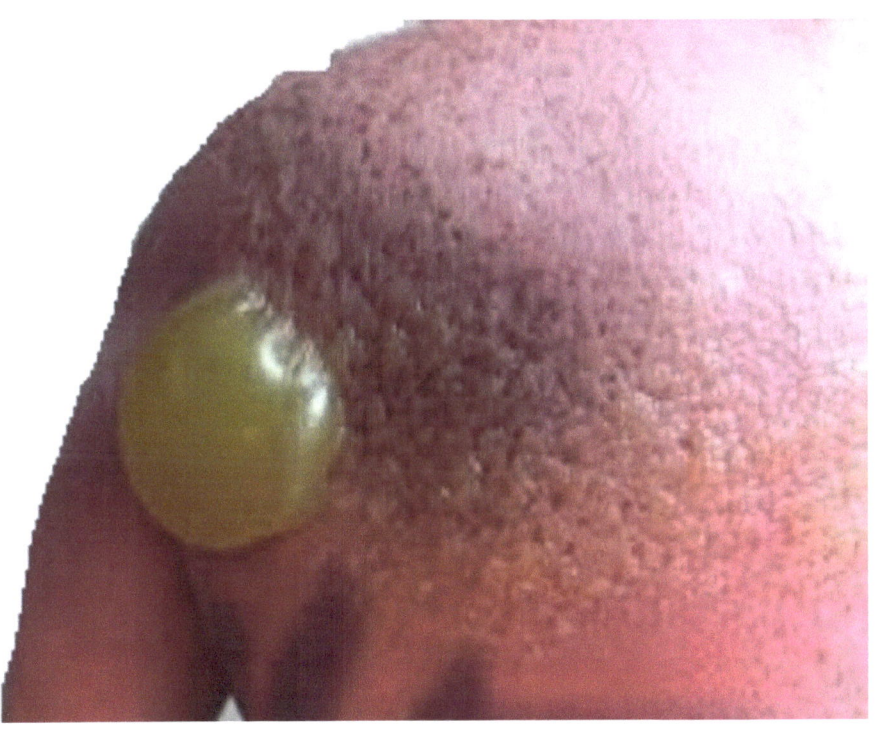

- Unwarranted exposure to sunlight can affect the eyes causing cataracts. UVR induced recurrence of conjunctivitis and corneal damage are also common. Moreover, vision impairment is very common when exposed to radiation especially during Solar eclipse.

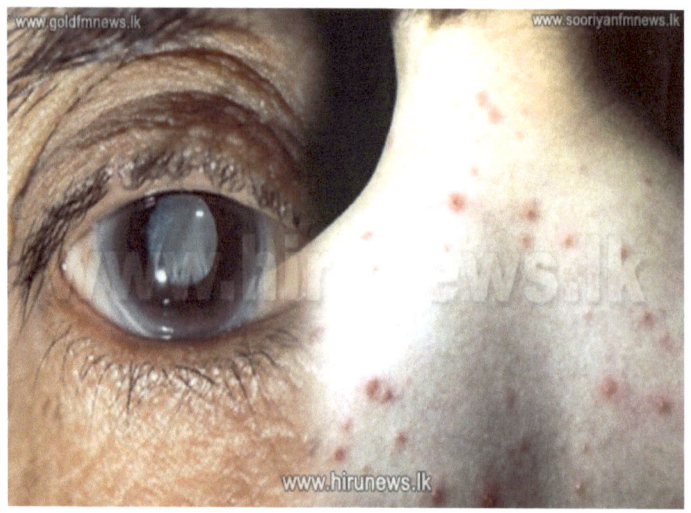

- UVA radioactivity enters deeply into the skin, where it can cause skin cancers. This radiation will result in the formation of DNA damaging molecules such as hydroxyl and oxygen radicals. Modern science recognizes that exposure to the ultraviolet radiation (UVR) results in helpful as well as harmful effects on human body.

- UV rays are considered as good disinfectants, which can protect us from various diseases. At the same time extreme UVR exposure can cause skin cancers like melanoma, basal cell carcinoma, and squamous cell carcinoma. Augmented dangers of various autoimmune illness, life-threatening cancers and musculoskeletal system disorders are also noticed.

# WHY SOLAR ENERGY?

Sunlight is considered to be the most beneficial form of energy. The solar radiation coming from space is being absorbed by atmospheric particles and dust. Part of the radiations is being reflected back to space by dust and clouds. Small percentage of radiations coming to earth is being absorbed by ocean water.

Vegetations on earth utilize the sunlight to prepare hydrocarbon by photosynthesis. Animal kingdom depends on this hydrocarbon for survival. Precisely the life on Earth is totally dependent on Sun directly or indirectly. Sunlight is a key factor that helps to purify air and to keep atmospheric air temperature above 15 degrees Celsius. Ultra violet rays also function as a good disinfecting agent of water and air.

Fossil fuels are formed from decayed vegetation which are millions of years old. Differential heating of atmospheric air is the reason behind the formation of air stream flow and wind energy. Wind is caused due to uneven heating of air by solar radiation.

Biomass is formed as a result of degradation of remains of plants and animal wastes which indirectly depends on solar energy as the vegetations derive their nutrition by photosynthesis in presence of sunlight.

Clouds are formed by the heat induced evaporation of sea water. Hydropower, derived from high altitude storage of rainwater which is made possible by adequate rain through the evaporation process. Every source of energy, either wind, biomass or hydropower depends on Sun either directly or indirectly. Hence Sun is the supreme provider of energy on earth.

Solar energy also finds application in filtering common salt from sea water by evaporation and to produce drinking water by desalinating sea water. Nowadays, solar application has grown beyond electricity production to hydrogen gas production. Its importance as clean and renewable alternate source of energy is widely accepted to combat climate change.

Artificial plant leaves are in experiment stage to split hydrogen and oxygen from water by mimicking the process of photosynthesis releasing no pollutants. Further effort is needed to advance the competence and cost effectiveness of these procedures for industrial use.

**Solar: A clean energy option**

Right from stone age, man has made step by step discovery of various uses of sun light which definitely made tremendous improvement in his lifestyle. Modern science unveils sunlight as primary energy source of the entire world, tuned to suit the necessity of all living creatures on Earth.

## Global warming

Promoting factors for the popularity of solar power installation are identified with the factors influencing global warming. With the present energy consumption rate of the world, oil reserve will last for 50 years, gas reserve will get depleted within 70 years and coal may last for 120 years. There is a great threat to energy reserves of the world with increasing energy demand and consumption rate.

Secondly the burning of fossil fuels causes global warming and air pollution. In this context of these crucial factors causing environmental catastrophe, promoting renewable energy especially solar energy is the way to combat this issue. Global community has introduced some restrictions on usage of fossil fuels and are encouraging promotion incentives for adapting renewable energy. Few international policies put forward are discussed below.

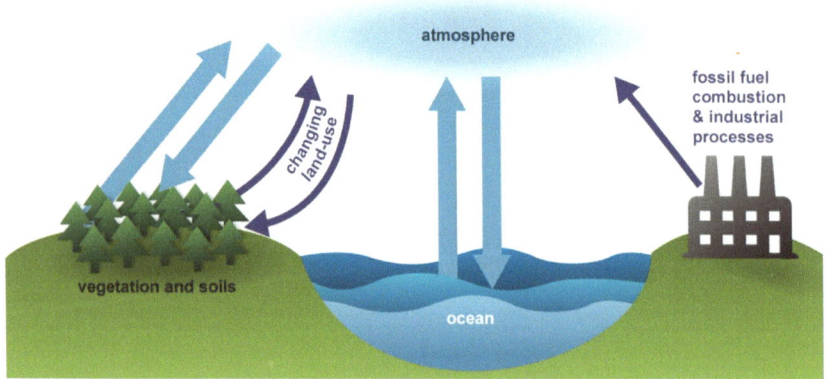

## Carbon credits

Carbon credits are certificates awarded to countries that have effectively reduced discharges of GHG thus reducing global warming. These can be effectively utilized by the industry world to prevent damaging carbon productions. Carbon credits generate market for reducing greenhouse gas discharges by giving a financial incentive.

Each carbon credit point indicates one ton of carbon dioxide ($CO_2$) either removed from the atmospheres or saved from being discharged. Carbon credits can be viewed as a means of authorizing the market to repair the environment. This can be bought and sold in worldwide marketplaces at prevailing market worth as this is the cost of contaminating the air.

## Clean Development Mechanism (CDM)

The carbon trade market was moulded with Clean Development Mechanism (CDM). The renewable energy players of developing countries are the main source of voluntary carbon credits in the world. At the same time this is interesting to note the fact that photovoltaic energy projects can nullify the emission of $CO_2$ at the rate of 1 kg per kWh of electricity.

**The Renewable Purchase Obligation (RPO)**

The Renewable Purchase Obligation (RPO) is functional in many countries to generate demand for renewable energy. With this legal norm, utilities are obliged for the acquisition of sure ratio of renewable power in the utility power mix.

The intentions of all above mechanism are to promote the growth and use of solar energy for multiple uses, with the eventual objective of making solar energy compete with conventional energy options. This will lessen the cost of solar power generation with large scale production and promote aggressive R&D activities in this sector. The promotion objectives of solar energy can be easily met by encouraging production of basic plant components at country level with viable financial assistance.

# SOLAR RADIATIONS

The Solar system is believed to be formed through a series of physical and chemical reactions. Some of the matter in space was accumulated at a centre point of that system drawn by gravitational force.

As a result, concentrated molecule formed the centre part and the dispersed matter stretched out to form the planets which go round the centre in separate orbits. These celestial systems started moving in a spherical path due to the dynamics of gravitation force acting between them.

During this entire process of formation, the matter accumulated at centre evolved to be Sun and became the hottest portion of the system radiating energy outwards. Scientists believe that this theory of fusion explains the formation of all stars which evolved in the universe. Modern science believes this theory, but many scientists differ in their opinion and the solar system, as we see today still remains a mystery.

SOLAR POWER

Sun attracts all the surface gases to the centre and all objects to its interior. The attractive forces are most felt at radioactive zone. The pressure and temperature at the centre are higher than the outer surface and its nucleus is the place where nuclear fusion takes place.

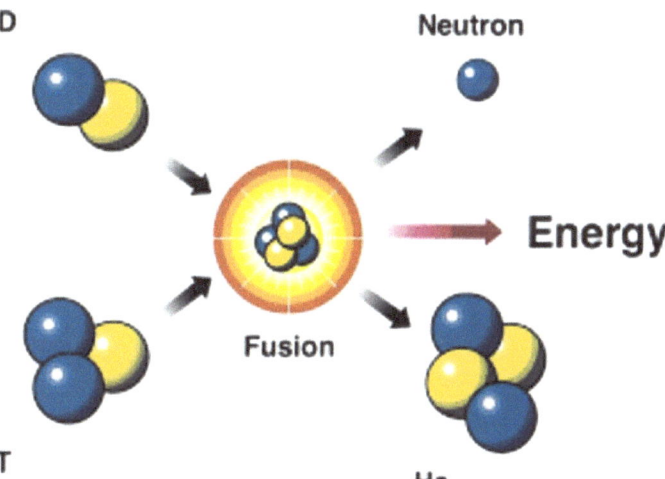

Energy created inside Sun is due to nuclear fusion reactions deep in its core, fusing hydrogen atoms into helium and releasing photons in the process. Nuclear fusion emits energy in the form of photons that travel towards the surface of Sun. very second, 700 million tons of hydrogen gases are converted to helium gas by nuclear fusion emitting quantum of energy in the form of photons.

The photons that are released during nuclear fusion bounce inside the core of Sun for thousands and thousands of years before defecating from the exterior to reach Earth. The moment at which all the hydrogen gas confined in the Sun is transformed to helium, the Sun will convert into a large meteor. Earth and the other planets will be pulled into it at the end. Different scientific thoughts predict the natural end of solar system by this way.

Each photon travels from nucleus to the surface of Sun, hitting gas particles during its course of journey. The colluding objects heats up the surrounding gases and the temperature at the surface is around 5800 Kelvin. The photon receiving atom heats up and passes the photon in random. When these photon release from the surface of Sun, the amount of energy emitted will be 63000 watts per square meter.

It takes almost 10,000 years for the radiation to reach the surface of Sun from the core of the Sun, but it takes only 8.3 minutes to reach Earth from the surface of the Sun. The Electromagnetic spectrum is generally divided into seven regions, depending on the decreasing wavelength and increasing energy and frequency. Following are the important types of radiation emitted by the Sun.

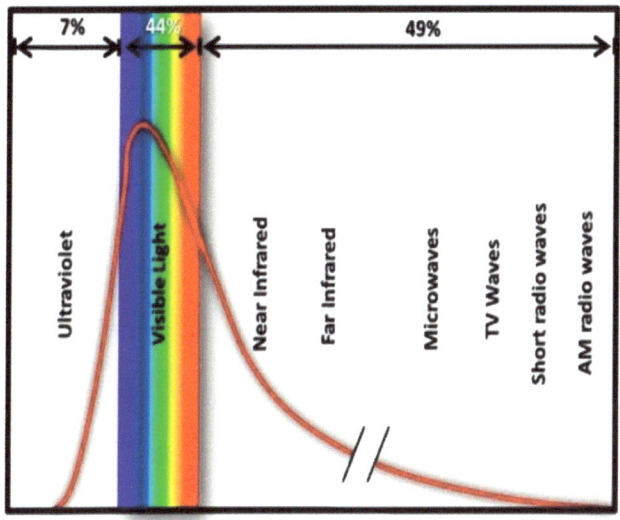

These radiations coming out from Sun surface can be categorized as electromagnetic radiations like

- Radio Waves
- Microwave radiations
- Infra-red rays
- Visible light
- Ultraviolet Rays
- X rays

In addition to electromagnetic radiations, particle radiations also radiating from the surface of Sun. These radiations include:

- Alpha radiation
- Beta radiation
- Proton radiation
- Neutron radiation

Among these radiations, the three relevant bands are the Infrared radiations (which make up to49.5 %), Visible light (42.5%), and Ultraviolet rays (8%). Ultraviolet rays are of three types-UVA, UVB, and UVC. UVA rays make upto 90% of the UV rays.

**Solar irradiance**

Sunlight falls on earth at every location at least for few days in a year. The amount of radiant energy emitted by the sun is called solar radiation, while Solar irradiation refers to the amount of radiation received from the sun per unit area. This is expressed as Watt per square meter.

Above the earth's atmosphere, Solar radiation has an intensity of approximately 1380watts per square meter. This value is called Solar constant. At our latitude, the value at the surface is approximately 1000watt per square meter on a clear day at noon in summer.

The quantity of solar radiation varies according to geographic position, time of day, season, landscape and weather conditions. When sunlight passes through the air, some of the radiations are absorbed, and few are reflected back to space. Water vapor, clouds, dust and other pollutants diffuse the solar radiation.

The Solar rays strike on earth surface at different angles, ranging from 0° (just above the horizon) to 90° (directly overhead).

When the rays are vertical, the Earth's surface acquires maximum energy. Figure shows the Solar radiation falling on earth.

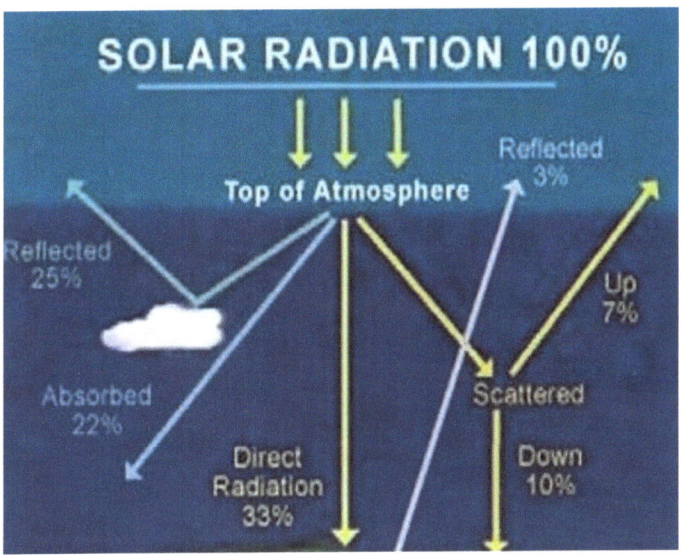

The more the rays are slanted, the longer they have to travel through the air, becoming more and more dispersed and diffused. That is why polar regions receives low level of solar energy.

The Earth revolves around the Sun in an oval path and is nearer to the Sun during some part of the year. When the Sun is nearer to the Earth, the Earth's surface accepts more solar energy. When Earth is nearer to Sun, it is summer in the southern hemisphere and winter in the northern hemisphere.

Oceans absorb the heat energy of the solar radiations and moderates the temperature of summer and winter. The 23.5° slant in the Earth's axis of rotation is a more important feature in defining the quantity of sunlight falling at a particular location.

Tilting results in longer days in the northern hemisphere from march to September longer days in the southern hemisphere during the September to march. Days and nights are both exactly 12 hours long on March 23 and September 22.

**Heat energy from solar radiations**

Solar energy, often called the solar reserve, is a universal term for the electromagnetic radiation released by the Sun. Solar radiation can be captured and converted into heat energy or electricity by suitable technologies. The technical and economic viability of this conversion depends on the availability of solar resource at a given location.

There are three primary technologies by which Solar energy

is captured or utilised. Photovoltaics (PV), which converts light to electricity directly, Concentrating Solar Power (CSP), which uses thermal energy to drive electric turbines, and Solar Heating and Cooling (SHC).

Heat energy from solar radiations can be tapped with help of lenses and mirrors. Lenses and mirrors are aligned in a spherical fashion to focus sunlight to a common point. Focussing of huge amount of solar radiation results in temperature rise at that point. Concentrated thermal solar plants deploy collectors to reflect sunlight of a wide area onto the receiver and thereby producing high temperatures in the order of 2000° C or more.

This accumulated heat energy can be drained by specifically designed pipes through which water is circulated. Heat transfer takes place and the generated water vapor is used to operate boiler plants of steam turbine power plant. The generated steam is made use in steam turbine to produce electricity.

# SOLAR ENERGY RESOURCE ASSESSMENT

Solar radiation is the key input and source for solar energy schemes. The accessibility of radiation straight away governs the returns of solar energy power plants. Data and information about this resource is vital to determine technical feasibility and economic viability. The direct normal irradiance is the quantity of solar radiation received directly from the Sun, falling on a plane perpendicular to the direction of the Sun.

It can be used for electricity generation via concentrating solar thermal power plants or concentrated PV. Direct irradiance has the benefit that it can be focused with mirrors to reach high temperature or high radiative flux. The drawback is that it is only available during clear sky. Therefore, energy schemes that use direct irradiance is only possible in sunny areas where clear sky environments are predominant.

**Global solar radiation.**

The solar radiation that reaches the Earth's surface without being diffused is called direct beam solar radiation. Dust particle in atmosphere reduce direct beam radiation significantly on clear dry days and marginally during thick, cloudy days. The amount of the diffused and direct solar radiation is termed as global solar radiation.

Before setting up a solar project, the amount of sunlight falling on specific locations at different periods of the year should be measured. Calculate the amount of sunlight falling on the regions of same latitude with similar climates. Measurements of solar energy are typically articulated as total radiation on a horizontal surface.

Radiation statistics for solar systems are often epitomized as kilowatt-hours per square meter ($kWh/m^2$). Direct evaluations of solar energy may also be articulated as watts per square meter ($W/m^2$).

**Resource Data**

Resource datasets provide information relating to solar irradiation of a specific region for plant design. There are free data sources as well as licensed data resources. The licenced data can be accessed by paying subscription.

These data sets either depend on ground-based measurements at meticulous meteorological stations or use administered satellite imagery. Quite a few data sets for meteorological analysis are available worldwide.

Actual ground data is not continuously accessible but derived data is obtainable from NASA, METEONORM and IMD. **NASA's** surface Meteorological and Solar Energy Program have satellite monthly data for a grid of 1°x1° (111 km) covering the whole world, using over 200 satellites and about 1100 ground stations. The Meteonorm Global Climatological Database and Synthetic Weather Generator contains a database of ground station measurements of irradiation, temperature etc.

Where a site is over 193km from the nearest measurement station it provides an output of climatologic averages estimated using interpolation algorithms. Where no radiation measurement station is available within 300km from the site, satellite information is used. If the site is between 50 and 300km from a measurement station a mixture of ground and satellite information is used. Monthly irradiance data is available from about 1700 stations as average for the period.

Geo Model Solar operates using high-resolution meteorological database. Solar radiation data is available for more regions of Europe, Africa, Asia, West Australia, and North and South America. Solar radiation is calculated by numerical models, which are parameterized by a set of inputs characterizing the cloud transmittance, state of the atmosphere and terrain conditions. Spatial resolution of satellite data used is about 4 x 5 km at mid-latitudes (3 km at sub- satellite point) and the time step is 15 and 30 minutes.

**Meteorological Department**

Meteorological Department (IMD) is established for meteorological observations and to provide current and forecast meteorological information for optimum operation of weather-sensitive activities like agriculture, irrigation, shipping, aviation and offshore oil explorations.

Meteorological Department will warn against severe weather conditions like cyclones, dust storms, heavy rains and snow fall. The details about cold and heat waves, which cause destruction of life, will be provided by meteorological statistics. This data is required for agriculture, water reserve management, industries, oil survey and other nation-building activities.

Assessment of solar energy potential for given location is the prime activity before starting the project installation. The plant design very strongly depends on the solar irradiance falling on that site. There are many viable tools and testing equipment available to assist in computing energy harvest on a given location.

## Testing equipment

Solar energy availability can be tested by different testing equipment. This is mandatory before solar plant installation. Testing the irradiance level and overall intensity of light on a planar surface is sufficient to calculate the approximate irradiance level of that area. Few testing equipment widely used for the above purpose are detailed below.

## Pyranometer

This is a type of actinometer used to measure solar irradiance on a planar surface. It is a sensor for measuring the solar radiation flux concentration, in watts per metre square, for a field of view of 180 degrees.

## Pyrheliometer

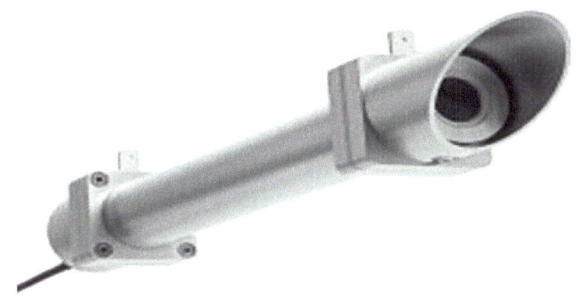

Pyrheliometer is to measure direct solar irradiance and is utilised for solar tracking by aiming the instrument at the Sun. Typical pyrheliometer, for measurement of direct solar radiation is shown in the figure.

**Lux Meter**

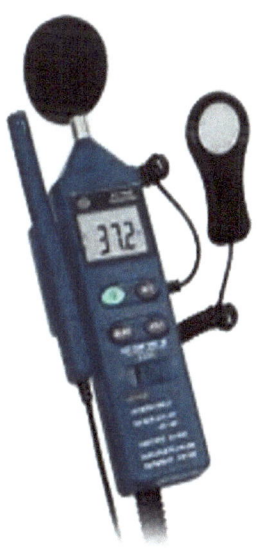

Lux meter is an equipment for measuring brightness in lux for the specific area in cd/m². Overall intensity of light within an environment for any given location can be measured. Photovoltaic cell is designed to capture light. Lux meter converts absorbed radiations to electrical c u r r e n t and the device will calculate the lux value of the captured light.

**Thermographic Camera**

Thermal imaging camera is an equipment that captures images using infrared radiation.

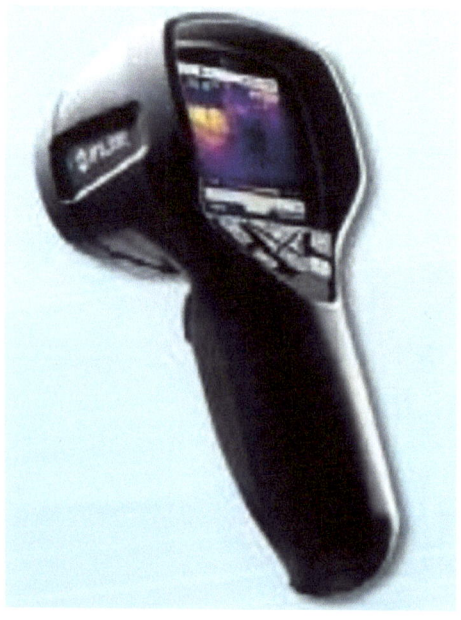

Its operation is similar to an ordinary camera that develop an image using visible light. Thermal imaging cameras operate in wavelengths in the range of 14,000 nm for thermography. They are also known as thermographic camera or infrared camera.

## Assessment of Annual Energy Generation

Energy yield from the proposed site can be easily calculated by using these tools during site visit. The average yearly energy yield can be estimated as follows:

$$E = P_{array} \times F_{temp} \times F_{man} \times F_{dirt} \times H_{tilt} \times K$$

where,
- $E$ = average yearly energy output of the PV array, in watt-hours
- $P_{array\_stc}$ = rated output power of the array under standard test conditions, in watts $F_{temp}$ = temperature de-rating factor, dimensionless
- $F_{man}$ = de-rating factor for manufacturing tolerance, dimensionless
- $F_{dirt}$ = de-rating factor for dirt, dimensionless
- $H_{tilt}$ = yearly irradiation value (kWh/m$^2$) for the selected site (allowing for tilt, orientation and shading)
- $K$ = efficiency of the subsystem including inverter and the switchboard.

The output of a PV module is expressed in watts at a given cell temperature. The output of a PV module can be reduced as a result of a build-up of dust on the surface of the module. The actual value of this de-rating will be dependent on the actual site but in some city locations this could be high due to the amount of pollution in the air. A satisfactory de-rating would be 5% for all sites and a derating of further 5% due to dirt can be included.

The average temperature of the cell within the PV module can be estimated by the following formula:

**T = T amp + 25°C**    where,

- T = average daily temperature, in degrees C and
- $T_{amb}$ = daytime average ambient temperature in degrees C.

Ground mounted stand-alone solar power systems are characteristically slanted at higher angles and the units have good airflow. On the other hand, rooftop grid models are exposed to higher temperatures. For on-grid-connect models, the actual cell temperature is calculated by the following formula:

**T = T amp + Tr**    Where,

- $T_r$ = effective temperature rise for specific type of installation.
- Parallel to roof (<150mm standoff): +35°C
- Rack-type mount (>150mm standoff): +30°C

Solar modules each have diverse temperature constants. The de-rating due to temperature will be dependent on the type of unit installed and the average ambient temperature for the location.

### Elevation (or altitude) angle (h)

Elevation angle is another determining factor of solar irradiance intensity. The elevation angle is the angular height of the Sun in the sky measured from the horizontal. The elevation is 0° at sunrise and 90° when the Sun is directly above. Elevation angle is shown in the figure.

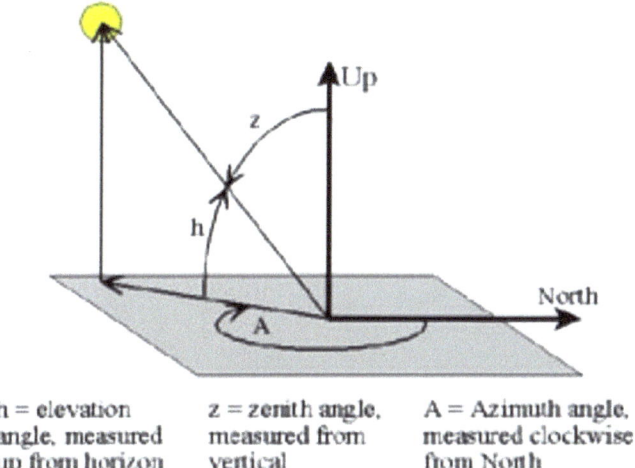

h = elevation angle, measured up from horizon

z = zenith angle, measured from vertical

A = Azimuth angle, measured clockwise from North

## Zenith Angle (z)

The zenith angle is the angle between the Sun and the vertical.

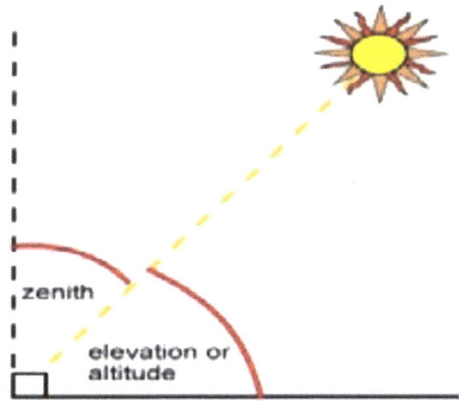

The zenith angle is similar to the elevation angle but it

is measured from the vertical rather than from the horizontal, thus making the zenith angle = 90° - elevation.

Z = 90° – h, Where, Z – Zenith angle, h – Elevation or altitude angle

**Azimuth Angle**

It is an angle measure on the horizontal plane clockwise from the north-pointing coordinate axis to the projection of the Sun's central ray. Azimuth angle is shown in the diagram.

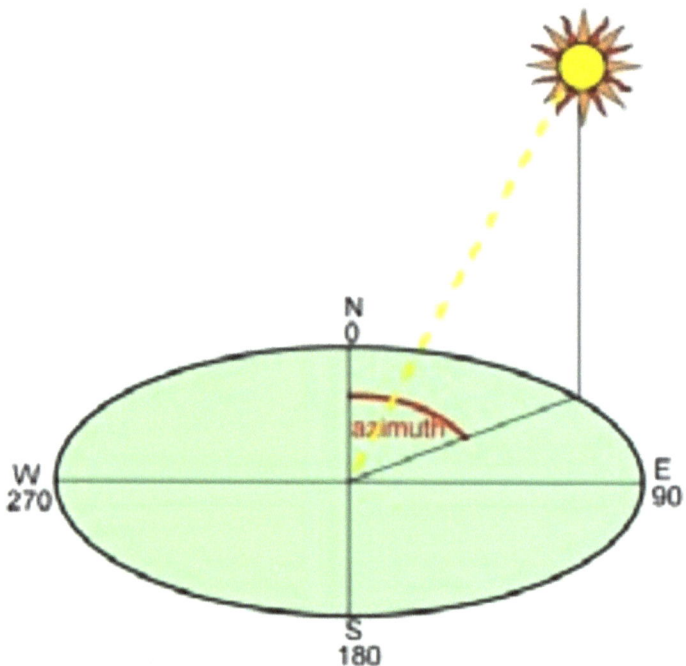

**Declination Angle**

The angle between the equator and a line drawn from the

centre of the Earth to the centre of the Sun is called declination angle. The declination angle, represented by "x", fluctuates seasonally due to the rotation of the Earth around the Sun and the tilt of the Earth on its axis of rotation. The declination angle can be computed by the equation: X = 23.45° sin [(360/365) * (d-81)],

Where, d is the day of the year.

**Tilt angle**

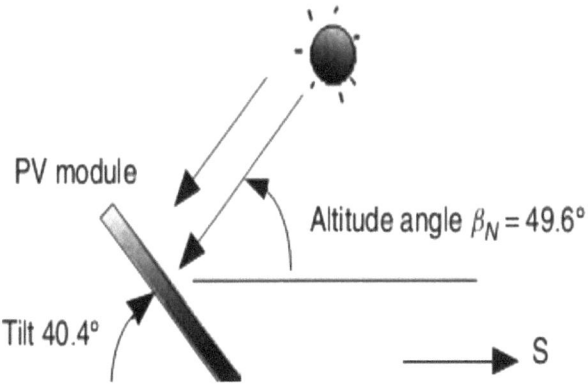

The angle, between array and horizontal surface is tilt and it is the angle which gives maximum solar irradiation.

Solar panels should always face true south in the northern hemisphere, or true north in the southern hemisphere. This shall give maximum incidence of Sunlight on the solar modules. Direction of solar panels should be oriented for capturing the most of solar radiations. Many other factors figure in the decision making for the best direction of panels.

Fixed panel or panels erected on an adjustable tilt mechanism can be decided for the system. Panels that track the movement of the Sun throughout the day can receive more energy than fixed panels. As the actual north and magnetic north are slightly different, correction regarding the slight difference in direction should be made while using compass.

Field experiments suggest that the tilt should be equal to the latitude, plus 15 degrees in winter or minus 15 degrees in summer. The graph below explains the impact of adjusting the tilt on energy harvest. The violet line is the amount of solar energy that we get each day if the panel is fixed for winter.

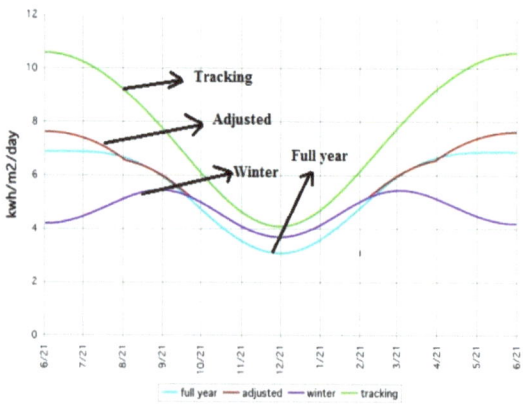

The greenish blue line displays the energy per day if the panel is static at the full year angle. The red line represents the energy yield by adjusting the tilt season wise. The green line displays the energy that we get from tracking, which always points the panel directly at the Sun.

# PHOTO VOLTAIC POWER

When the incident electromagnetic radiations or photons strikes on semiconductor junction, electrons gain energy and whenever the energy acquired by the electrons exceeds threshold limit it starts moving away from nucleus. This energized electron is generally known as photo electrons and phenomenon of emitting electrons from a semiconductor material by solar radiation is known as photo-electric effect. The basic working principle of a solar cell is based on this photoelectric effect.

**Solar cell**

The solar cell is a device that produce electricity from solar radiation due to its semiconductor features. When light strikes on a solar cell, it may be reflected, absorbed, or passes right through it. But only the absorbed light generates electricity.

The solar cell is made with semiconductor material such as silicon or selenium. When the two differently doped P type and N type semiconductors are joined there exist a natural electric potential, which stimulate the electrons to flow from one region to another region once exposed to sunlight.

These materials absorb sunlight and eject the loosely bounded outer orbit electrons. The conversion of solar energy to electrical energy is achieved by striking photons on solar cells and is proportional to the intensity of radiations falling on solar junction.

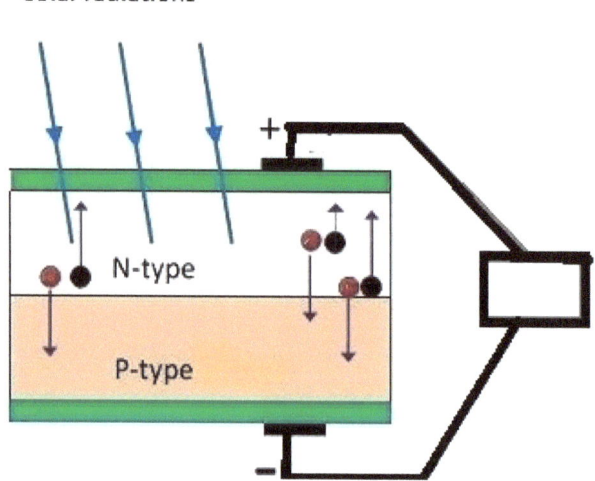

Photo electric cells or solar cells are generally manufactured with semiconductor materials like silicon, selenium arsenide or gallium. Most common semiconductor materials used for photo electric cell are silicon and selenium.

The exterior of the cell is made of the thin film of the p-type substance for the easy entry of solar radiation into the material. Positive and negative terminals of the cell are moulded over P type and N type to form a cell. Sandwiched junction of semiconductor material forms a cell configuration, which can produce maximum of two watts of electricity. But super power plants can be set up with millions of such cells arranged in solar panel arrays.

**Solar Panels or Module**

To improve the efficiency of energy conversion and to increase amount of electricity generation, large number of individual PV cells are unified together in a wrapped, weatherproof platform. This is called a panel.

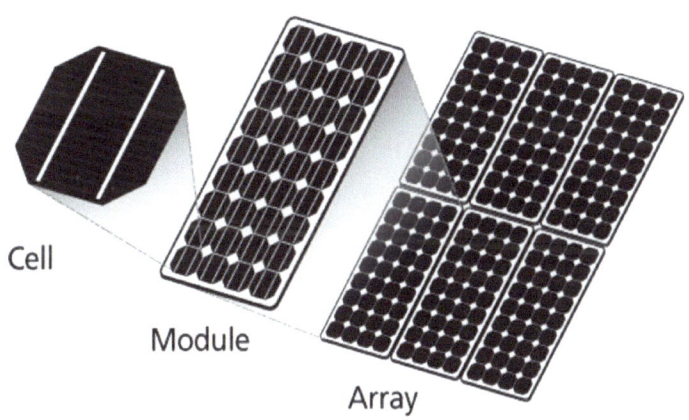

In order to achieve the predicted power output, solar panels are inter connected in different combination of series and parallel connection and what is called a PV array. Series grouping of modules is also known as a string.

**Power output from solar system**

The solar energy conversion ability and efficiency of a PV module is described by its current and voltage (I-V) characteristics. Energy output performance and solar efficiency can be determined by knowing the electrical I-V characteristics (more importantly Pmax) of a solar cell, or panel. The current-voltage (I-V) characteristics of a silicon PV cell operating under standard circumstances is shown in figure.

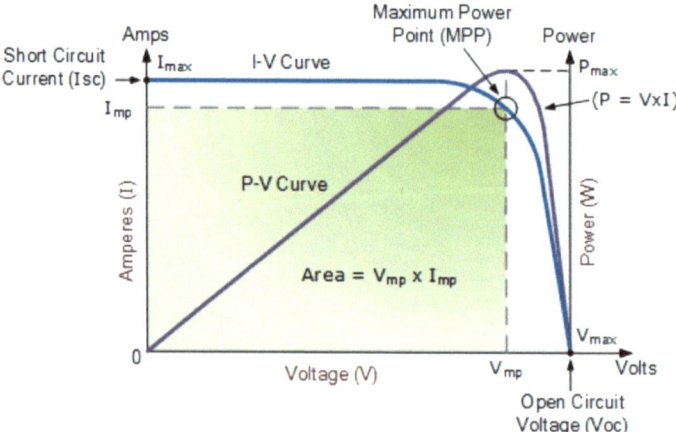

The power produced by a solar cell is the product of current and voltage (I x V). If the multiplication is done, point for point, for all voltages from short-circuit to open-circuit conditions, the power curve can be drawn for a given solar radiation level.

In solar cell open-circuited condition, which is not connected to any external load, the current will be at its minimum (zero) and the voltage across the cell is at its maximum. This is known as the solar cells open circuit voltage or **Voc**.

When the solar cell is short circuited, which means the positive and negative leads connected together, the voltage across the cell is at its minimum (zero) and the current flowing out of the cell ranges its maximum. This is known as the solar cells short circuit current, or Isc. It can be seen in the diagram that the top right area of the shaded rectangle points at which the cell generates maximum electrical power is the "maximum power point" or MPP.

It is presumed that the ideal operation of a photovoltaic cell system is expected to be at the maximum power point. The maximum power point (MPP) of a solar cell is positioned near the bend in the I-V characteristics curve. The corresponding values of **Vmp** and **Imp** can be estimated from the open circuit voltage and the short circuit current.

The **fill factor** is another term used to explain the efficiency of solar plants. It represents the relationship between the maximum power that the array can actually provide under normal operating conditions and is the product of the open-circuit voltage times the short-circuit current, (Voc x Isc). This fill factor value gives an idea of the quality of the array and the closer the fill factor to 1 (unity), the more power the array can provide. Typical values are between 0.8 and 0.9.

**Combinations of panels**

Series or parallel combinations, or both types of the photovoltaic panels can be connected together to increase the voltage or current capacity of the solar array. If the solar panel are connected together in a series combination, then the voltage increases and if connected together in parallel then the current increases.

In both cases the solar power output in Watts, generated by these different photovoltaic combinations will still be the product of the voltage times the current, (P = V x I). The upper right-hand corner of shaded portion shown in graph will always be the maximum power point (MPP) of the array.

**Effect of temperature on V-I characteristics**

Ambient temperature has significant effect on power production from a solar cell. With the increase in ambient temperature, Voc of the cell reduces, whereas Isc remains constant.

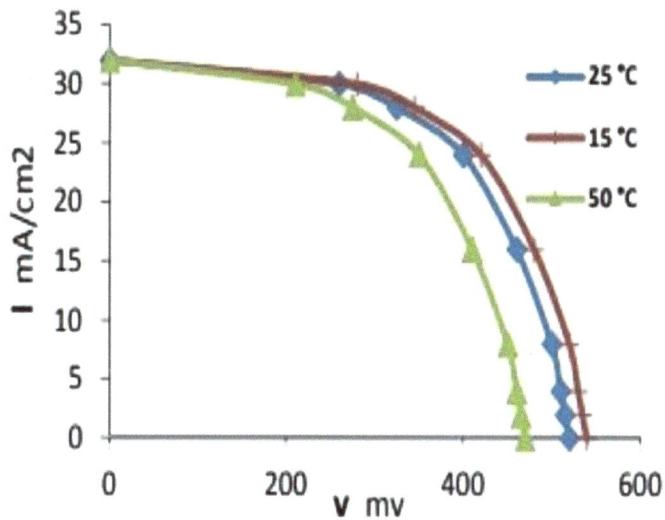

## Effect of irradiance on

## V-I Characteristics

As the solar irradiance increases, Isc increases radically, whereas Voc reduces by small amount (almost constant). This shown in the diagram.

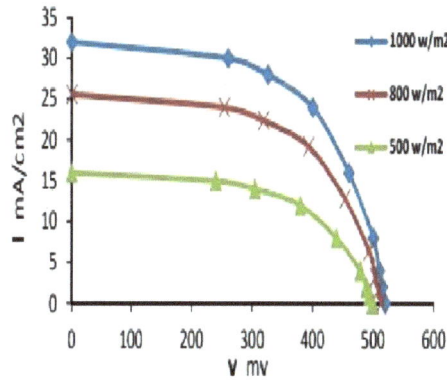

## System Stability – Cloud Effect

The system stability is affected by cloud passing. This will reduce the power output. When cloud passes the output, waveform is distorted and system stability is badly affected.

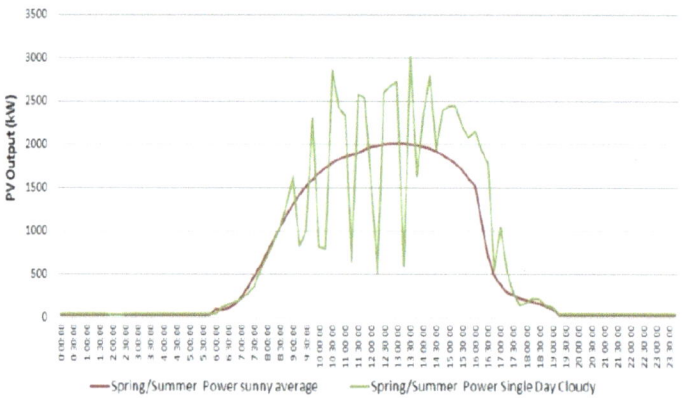

## Effect of shadow on Solar PV Modules

Shadow falling on solar cells will reduce the total output by reducing the energy input to the cell and by increasing energy losses in the shaded cells.

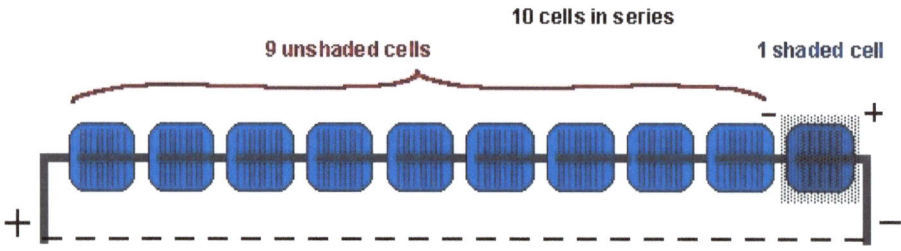

Shading effects are more significant when shaded cells get reverse biased. Temperature raise is a common phenomenon seen on shaded cells which may be leading to increase in its internal resistance.

As a result of increased temperature, the cell current gets concentrated in an increasingly small region of the cell, producing the hot spot. This can damage the cell and eventually cause entire module failure. Partial shading will adversely affect the performance of a series connected string of solar system if all its cells are not equally illuminated.

In a large area solar array, it is likely that shadow may cause due to tree leaves falling over it, birds or bird litters on the array or shade of a neighbouring plants. In a series connected string of cells, all the cells transmit the same current. Even though a few cells beneath shade yield less photon current but these cells are also forced to transmit the same current as the other fully well-lit cells. The sheltered cells may get reverse biased, acting as loads, demanding power from fully illuminated cells.

If the system is not properly protected, hot-spot problem can arise eventually, the panels can be permanently damaged. Shading of panels due to nearby roof is shown in figure.

**By pass Diode**

If many cells are connected in series, screening of individual cells can lead to the ruin of the sheltered cells or of the lamination materials. This may end up with panel (Module) swelling.

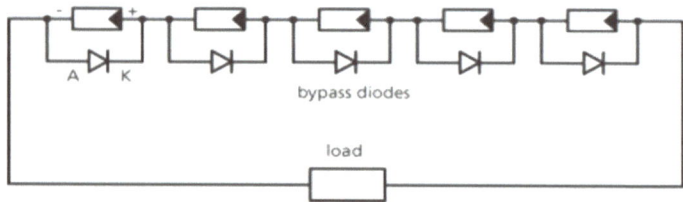

To overcome such critical situations, a new device named

bypass diode is connected anti-parallel to the solar cells as shown in the figure. Accordingly, greater voltage differences cannot get up in the reverse current direction of the solar cells. In practice, it is adequate to connect one bypass diode for every 15-20 cells.

Bypass diodes also allow current to flow through the PV module when it is moderately shaded, even if at a reduced voltage and power. Bypass diodes do not cause any losses, because under normal operation, current does not flow through them.

# CRYSTALLINE TECHNOLOGY

The most commonly used material in PV industry is crystalline silicon. The first available market driven technology of solar panels is crystalline silicon-based technology referred to as first generation with wafer based, carbon silicon cells. Wafer based carbon silicon PV cells lead the present commercial market place.

The engineering procedure of wafer-based silicon PV modules comprises various steps such as wafer construction, cell construction and module assembly. The voltage output from a photo voltaic cell is less than 1 Volt. But required voltage can be made available by series and parallel combination of cells. Crystalline silicon cells are categorized into different types based on how the silicon wafers are made.

They are:

- Monocrystalline (Mono c-Si)
- Polycrystalline (Poly c-Si),
- EFG ribbon silicon and silicon sheet-defined film growth (EFG ribbon- c-Si)

## Monocrystalline Cell Technology

Majority of the solar cells made by using crystalline silicon in electronics industry. These are widely used because of its wide availability as a raw material. Based on the crystalline structure this first-generation technology is further classified into monocrystalline and poly crystalline respectively.

In the case of mono crystal, the entire cell is cut from a semiconductor material as a single unit. But poly crystalline cells are obtained from different parts which have many sides. The raw material used for monocrystalline cell is silicon oxide ($SiO_2$), available in the form of sand. The raw sand is transformed into metallurgical grade silicon in a furnace through a reduction process using coal.

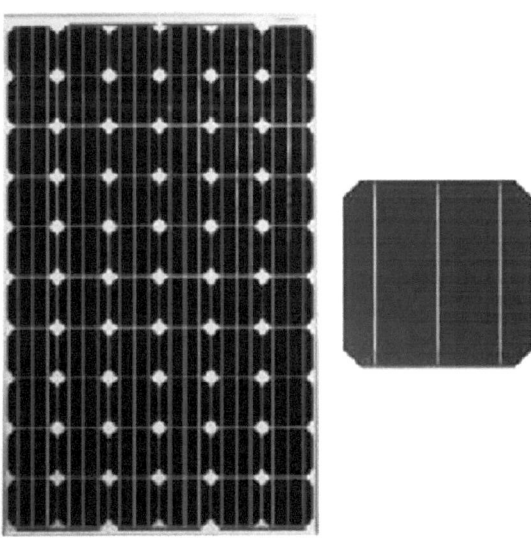

Vapor deposition is the chemical process in which silicon nuggets are formed. This is achieved by allowing $SiO_2$ to react with hydrogen and chlorine.

The silicon bits are liquefied in a container by using the Czochralski method, in a crystal puller. The silicon ingots are cut by wire saw into wafers of 200–250 micron thick which are very expensive. In the crystal drawing process, precise temperature control is required.

The temperature limits should be within ± 0.1°C of a melt at 1400°C in passive atmosphere. After the labour-intensive process of cutting, grinding and polishing, the wafers are placed through a gassy dispersion process for bonding with an added material.

The efficiencies in the order of 23% have been obtained with silicon cells under test conditions but on marketable basis it is around 15%. Best profitable modules presently on market have efficiencies of around 20%. China, India, United States of America and Canada are the leading countries engineering monocrystalline modules.

**Poly crystalline cell technology**

Many companies have industrialized devices to produce polycrystalline silicon blocks which can be used for solar cell production. Low-level purity is required for producing semiconductors for solar PV cells. Hence silicon purification process and crystal growing process are avoided in polycrystalline process. Scrap silicon of low purity is used here for a process to produce slabs of silicon which is then sliced into wafers like monocrystalline.

The efficiencies of polycrystalline solar cells are lower than monocrystalline at the commercial level. But poly crystalline panels are having the advantages of reduced wastage of silicon and reduced production cost. India, China and Canada are the leading manufactures of poly crystalline cells. Figure of polycrystalline silicon module is shown below.

## Thin film based- 2nd Generation

Thin-film photovoltaic cell is made by placing one or more thin films of photovoltaic substance on a substrate. The thickness range of such a coating is wide and varies from a few nanometres to tens of micrometres. The latest thin-film technologies have reduced the quantity of semiconductor material required in producing a solar cell to minimum level.

Compared with other crystalline technology, the efficiency of these cells is very low, but cost of production is low. Thin-film photovoltaic cells have become more popular compared to crystalline silicon due to lower costs.

The thin film modules can be further classified based on the type of semiconductor layer.

- Amorphous silicon (a-Si)
- Cadmium Telluride (CdTe)
- Copper-Indium/Gallium-Diselenide/Disulphide (CIS, CIGS)
- Multi-junction cells (a-Si/m-Si)

**Amorphous Silicon Solar Cells**

Unlike in crystalline cells, atoms are organized in a random manner in amorphous silicon module. It can absorb almost 40 times more sunlit than monocrystalline silicon. Therefore, solar cells made of amorphous silicon could be very thin and thus raw material cost required is very less. The important advantages of amorphous silicon units are as follows:

- High optical captivation
- Higher band gap
- Less material ingesting
- Low energy feeding during production
- Option of mechanization in the manufacturing process

Major manufacturers of a-Si thin-film modules are Sharp Solar, Japan and Next Power, Taiwan.

**Cadmium telluride solar cell**

In cadmium telluride solar cell, a thin film made up of cadmium telluride (CdTe) is used as semiconductor layer to absorb and alter sunlight into electricity. Cadmium, being a toxic material present in the cells, utmost care should be taken during construction and use of these types of cells.

In normal case cadmium release is negligible as fire occurrence on residential roofs are unlikely. A square meter of CdTe comprises roughly the same amount of cadmium used in a single cell of Nickel-cadmium battery. The efficiency of cadmium telluride cell is roughly 6%.

1972 marked the beginning of solar cell based on cadmium. Afterwards substantial development has been made in the cell and highest efficiency of 16.5% has been achieved in recent times.

Cadmium telluride is highly acceptable to solar industry due to its optimum optical properties of energy absorption and conversion. Cadmium telluride is considered as straight band gap semiconductor having high absorption efficiency with low-cost production techniques. First Solar, USA, is the main producer of cadmium telluride modules in the world.

**Copper-Indium Selenide**

Copper-Indium Selenide solar cells has achieved conversion efficiency greater than 14%. The manufacturing costs of CIS solar cells are high compared to amorphous silicon technology solar cells. Advanced research work is required for cost-effective manufacturing of CIS cells.

Gallium can be used as a replacement material for indium due to its free availability. Gallium is an addictive material for increasing the optical band gap of the CIGS layer as compared to pure CIS. This will increase the open-circuit voltage and will decrease the short circuit current. Solar Frontier, of Japan, is the main producer of CIS units in the world. Concentrating PV (CPV) and organic PV cells are innovative technologies and new concepts under progress, but not yet popular.

Comparison between Monocrystalline, Polycrystalline and thin film module are shown in the table.

| Mono-Si Panels | Poly-Si Panels | Thin Film Panels |
|---|---|---|
| 1. Most efficient with max. efficiency of 21%. | 1. Less efficient with efficiency of 16% (max.) | 1. Least efficient with max. efficiency of 12%. |
| 2. Manufactured from single Si crystal. | 2. Manufactured by fusing different crystals of Si. | 2. Manufactured by depositing 1 or more layers of PV material on substrate. |
| 3. Performance best at standard temperature. | 3. Performance best at moderately high temperature. | 3. Performance best at high temperatures. |
| 4. Requires least area for a given power. | 4. Requires less area for a given power. | 4. Requires large area for a given power. |
| 5. Large amount of Si hence, high embodied energy. | 5. Large amount of Si hence, high Embodied energy. | 4. Low amount of Si used hence, low embodied energy. |
| 6. Performance degrades in low-sunlight conditions. | 6. Performance degrades in low-sunlight conditions. | 5. Performance less affected by low-sunlight conditions. |

# COMPONENTS OF SOLAR PLANTS

A Solar photovoltaic plant is an electricity generating unit using photovoltaic modules. Excluding the conventional prime movers, solar plant is a basic generating plant with solar module as energy producers.

This plant is similar to any other type of conventional plants such as hydro, biomass or thermal when the electric equipment connected are compared. In solar plants transformers, HT/LT panels, electric cables, switch yard equipment and metering equipment, lightning protection and earthing equipment are used.

The important advantage of solar plant is that the module needs little maintenance as there are no moving parts. On the other hand, all other electrical and switching equipment necessitate suitable and specified maintenance and substitutions as per requirement. A solar power plant has various equipment for generating electricity. The electric power produced by solar modules goes through a series of changes before it reaches the grid. Those conversions specifically include adjustments of current, voltage and DC-AC conversion generally known as power conditioning.

Power conditioning is an important function of any on grid solar plant, which ensures that the energy generated can be effectively and safely delivered to consumers. To achieve the proper power conditioning, we need a number of dedicated components in addition to solar panels.

Photovoltaic plants comprise a large number of subordinate equipment, which serves to balance the system and to make it sustainably operational. The energy flow from the solar plant is through a variety of power conditioning devices, which are connected by wire network and related hardware.

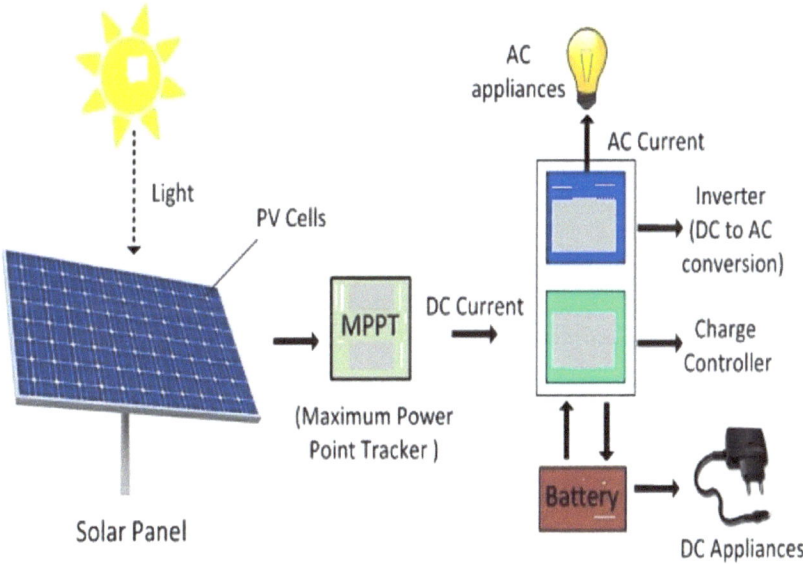

The additional mechanisms include transformers, connector boxes, inverters, controllers, energy storage devices, switches, monitoring devices, charge regulators and other accessories. Grounding and bonding of related DC and AC circuits is important to maintain system integrity.

These equipment help to condition the generated electric power for various applications. PV systems are characteristically modular in design, so that extra sections can be added to the plant or detached for maintenance without any major interference to its structure.

**On-Grid Inverter:**

On-grid inverter converts the DC power to AC power and is one of the indispensable components of PV system to inter connect with the utility grid. Various type of inverters available in the market with different power rating starting from small kVA to larger kVA. The modern inverters are incorporated with MPPT for input Vdc range.

**Charge Controller**

This device is to control the voltage drawn from the system and to protect the battery system from overcharging. Charge controllers or regulators monitor and control the flow of electricity between the solar modules and loads.

The suitable charge control algorithm and charging currents

should be harmonized for energy storage devices used in the system. The foremost purpose of a charge controller is to safeguard batteries from damage and to avert overcharging or unwarranted discharging. Characteristically, these strategies operate in the switch on / switch off mode. Charge controllers also contribute in voltage alteration and maximum power tracking.

## Maximum Power Point Tracker

The location of the Sun with respect to the installed solar plant varies within the day and season due to the movement of the earth. So, for absorption of maximum light, the panel should be oriented against the Sun always. Maximum power point tracker is an algorithm which helps the solar cells for maximum power generation.

Solar plants use maximum power point tracking (MPPT) to achieve the maximum possible power from the PV system. The injected current must be controlled such way that it is balanced in all available phases.

Although it utilizes measured values of grid voltage and frequency, it cannot control the line voltage or frequency at the point of interconnection to the grid.

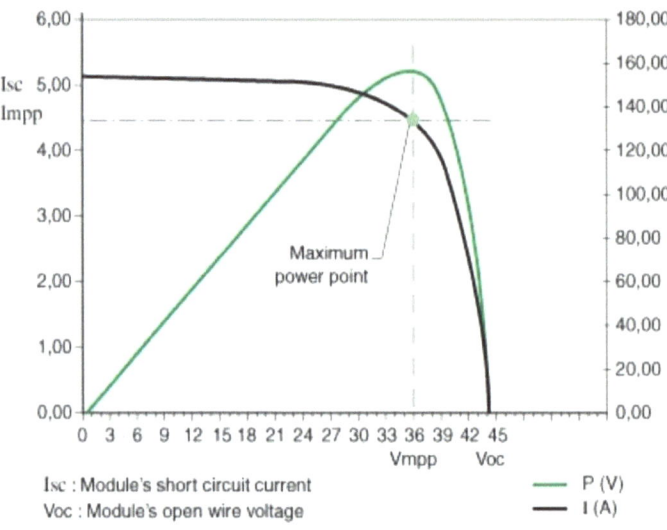

Isc : Module's short circuit current
Voc : Module's open wire voltage

The maximum point of power of the PV modules matches to the point of tangency between the I-V characteristic for a given value of solar radiation and the hyperbola of equation VI = const. Figure shows the maximum power point tracking of PV module.

The MPPT systems commercially used to identify the maximum power point on the characteristic curve of the PV module. This is achieved by small changes of loads at regular intervals which control deviations of the voltage current values and assess if the new product I-V is higher or lower than the previous one.

Due to the characteristics of the mandatory acts the inverters for stand-alone plants and for grid-connected plants shall have different features. For stand-alone plants the inverters shall be able to supply a constant AC voltage by varying the production of the generator. But for grid-connected plants the inverters shall reproduce the network voltage and at the same time optimize and maximize the energy output of the PV panels.

**AC Main Panel**

Main panel comes into picture before the electrical system is integrated to power grid. This generally consists of electro mechanical devices that are used to disconnect the photovoltaic system from the electric grid.

AC disconnect switch in AC main panel is to separate the on-grid DC-AC inverter from the electricity utility grid. Output currents of the inverters have to be taken into consideration while sizing the AC disconnect switch.

**Net Meter**

In solar plants, the excess electricity generated is sold to the utility by net meter arrangements. Net meter is installed for the purpose of measuring and monitoring the inflow and outflow of electricity between the solar plant and electricity utility grid.

**Junction Box**

The junction boxes were used mainly for the inter connection to power converter and for the junction to use the bypass diodes. All the PV strings are joined together at solar PV enclose where this junction is used for the bypass diodes allowing the power flow only from solar panel to the utility system.

**Electricity Grid**

The electrical network bridges the link between solar plant

and the consumers and delivers electric power as per consumer requirement. It is the network interconnecting the consumers and energy producers. It acts as an interface between power generation plant, power transmission network and distribution network. It is to transmit and distribute electric power from generating stations to the end user.

**Batteries**

Batteries are used in many types of PV systems to supply power at low Sunshine conditions. Additionally, batteries are required in solar systems because of the fluctuating nature of the PV output. The battery capacity is selected according to the load.

Batteries are usually connected in parallel to match higher capacity. There are several types of batteries commercially available for solar applications, including lead-acid, nickel-cadmium, nickel hydride, and lithium-ion. The foremost requirement for the batteries that are used as energy storage for solar systems is that they must be able to go through deep charging and discharging cycles without too much degradation.

# SOLAR INVERTOR

Inverters perform a vital role in solar energy conversion. Alternating current is the typical power used by all commercial and home appliances. The inverter does the purpose of converting the direct current produced from solar system into the alternating current and vice versa as most of house hold equipment works on alternating current.

The elementary function of an inverter is to "invert" the direct current (DC) output of the photovoltaic (PV) system into alternating current (AC). In practice inverters acts as a "gateway" between the photovoltaic (PV) system and the utility grid.

As technology advances, in addition to the mere job of DC to AC conversion, it also delivers other facilities to ensure the optimal performance level of entire system. Modern inverters are equipped with monitoring devices, advanced utility control mechanisms, software applications. Inverters are broadly classified in to grid tied invertors and off grid invertors.

**Grid tied inverter**

Grid tied PV systems consist of solar panels and a grid-tie inverter, with no batteries. The solar inverter is placed between the solar array and the utility grid. Solar panel of a specific rating produce a precise amount of energy which is then conditioned and managed by using inverters to deliver an AC power to the grid.

The solar panels supply electricity to grid-tie inverter which changes the DC power coming from the solar panels directly into AC power to tie with the utility grid.

If any consumer generates more solar power than consumption, then the extra power is pumped to the utility grid. Any power supplied by the consumer to utility is deducted from consumer consumption bill by the power company.

The inverter must monitor the grid voltage, waveform, frequency and other dynamic variants. The main reason for monitoring the grid is for healthy operation of the system and for synchronization with the grid waveform. One of the reasons for monitoring is, if the grid is off, and inverter connected to a faulty power line, to disconnect the invertor automatically in accordance to the safety rules.

The invertor is supposed to produce a slightly high voltage than the grid voltage itself in order to get better sinusoidal waveform with less harmonics.

A high-quality modern on-line grid inverter has a very good power factor. They produce AC output voltage and current which are perfectly lined up wave forms, and its phase angle is very less deviated with the utility grid. Figure shows a simple grid tied invertor.

It is important to keep the grid voltage within allowed limitations when more numbers of renewable plants are connected to local grid. Voltage levels might rise much higher than desired, during noon time with solar plants.

Grid-tie inverters are also planned for speedy disconnection from the grid if the utility grid goes down. When there is shutdown, the grid tie inverter will switch off to prevent utility worker from getting harmed while attending to the internal fault.

**Inverter Circuit**

As mentioned earlier, invertor does the job of converting DC to AC. To generate three phase outputs, 50 Hz transformer-based central inverters typically use a 3-phase bridge, whereas for single phase output, single phase bridge is required. The bridge shown below is with a 4-switch design. Typical inverter circuitry (H-Bridge) is shown below.

The DC voltage is inverted from positive to negative by alternately closing the top left and bottom right switches, then the top right and bottom left switches. By this operation rectangular AC waveform is created. The role of the filter circuit after the H-bridge is to smoothen the curve and change the magnitude of the obtained waveform so as to resemble or become identical to the sine wave.

SOLAR POWER

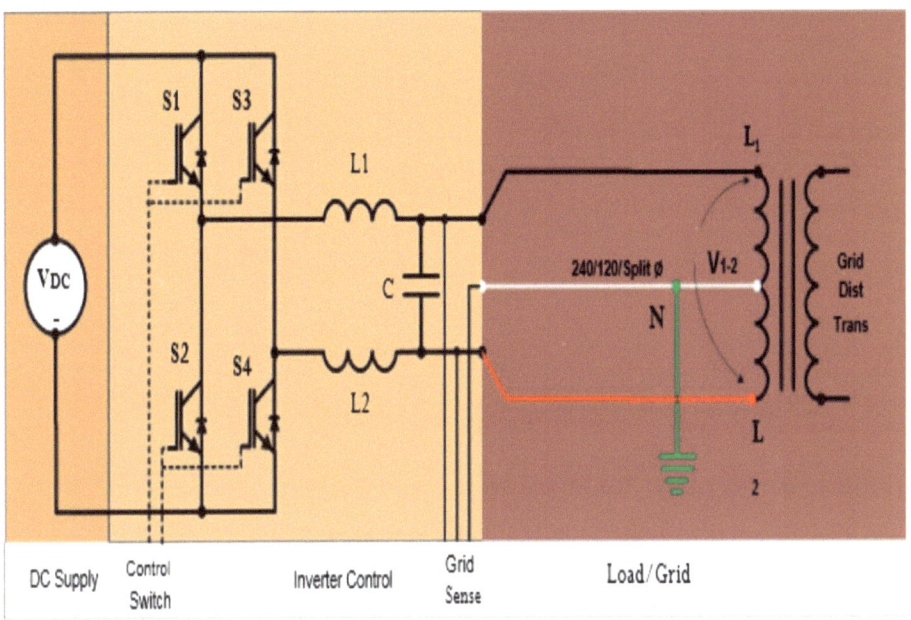

The filter includes the inductor, the capacitor and the transformer. These mechanisms filter the wave shapes resulting from the PWM switching, flattening out the sine waves, and bring AC voltages to the correct levels for grid interconnection.

The magnetics also deliver separation between the DC circuits and the AC grid. The figure represents wave forms of pulse width modulation bridge output.

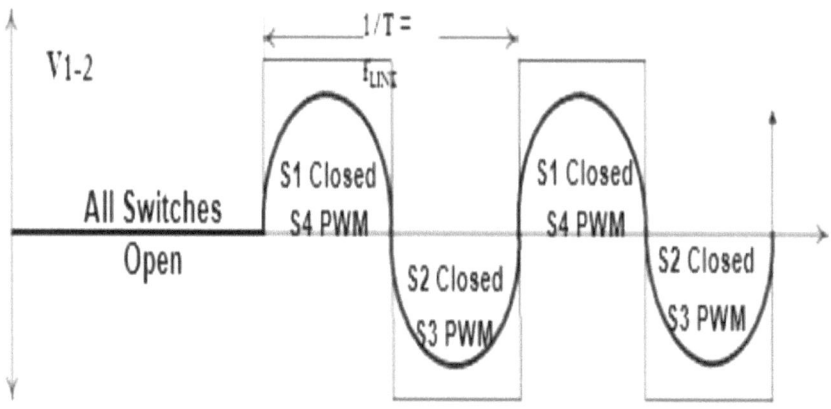

The AC waveform entering the filter circuit is raw and irregular but on leaving the device, it is a clean 180 Vac sine wave. Because 180 Vac cannot be directly connected to the utility grid, it goes through a matching transformer. The resulting smooth, sinusoidal 240 V AC is connected to Grid. Grid synchronisation process is made possible by grid sensing feedback.

Grid voltage information is provided to the inverter's digital signal processor or microcontroller, the device that controls the H-bridge. The inverter's DC input bus voltage needs to be greater than the peak of the AC voltage on the primary side of the transformer.

In order to continue this relationship at all times, an additional control and safety margin is required. With a minimum PV input voltage of 250 Vdc, for example, the highest amplitude of AC sine wave you can create is about 180 Vac. The most significant use of capacitors in the inverter power stage is for filtering ripple currents on DC lines.

Ripple is an unwanted phenomenon caused by power semiconductor switching. Capacitors are also used to keep the DC bus voltage stable and minimize resistive losses over the DC wiring between the PV array and the inverter, since the resulting current from the PV system to the inverter DC bus is constant.

The L-C filter inside the inverter will do the conditioning of power supply output. The control system of the inverter that converts direct current to alternating current controls the quality of the output power to be delivered to the grid.

Grid line voltage and frequency measurements are used by inverters that injects available energy from a PV array into the connected grid to synchronize to its grid connection. It operates as a single phase to three phase current source.

Static switches such as transistors are used for controlling the opening-closing signal which will result in an output square waveform. This is converted to a more sinusoidal wave form by using Pulse Width Modulation (PWM) technique. PWM technique permits required regulation on the frequency as well as on the rms value of the output waveform.

**Pulse Width Modulation**

Pulse width modulation is the technique of regulating the width of the pulse train in direct proportion to a small control signal. The healthier the control voltage, the broader the subsequent pulse result. By using a sine wave of predicted frequency as the control voltage for a PWM circuit, it is possible to produce a high-power waveform with normal voltage fluctuates in sinusoidal way.

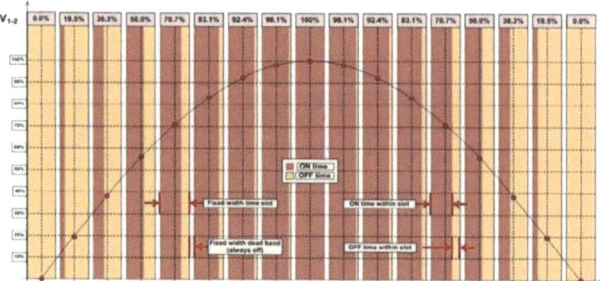

The below picture shows the PWM output over the positive half cycle. Waveform is separated into a number of small equal time segments determined by PWM switching frequency. Each segment is turned on for a given duty cycle depending on voltage of the grid.

**Micro-inverters**

A solar micro-inverter is an equipment used in photovoltaic that changes direct current (DC) generated by a single solar module to alternating current (AC). The output from several micro inverters is collected and supplied to the utility grid.

Micro inverters have numerous merits over conventional inverters. The main advantage is that small amounts of shading, debris or snow lines on any one solar module, or even a complete module failure, do not unreasonably reduce the output of the entire array.

Each micro inverter harvests optimum power by performing maximum power point tracking (MPPT) for its connected module. Simplicity in system design, lower wiring expenses, and added safety are other factors merited with the micro inverter.

Micro inverters are minor inverters rated to handle the output of a single panel. Modern grid tie panels are usually rated between 250W and 500W, but rarely produce this in practice, so micro inverters are typically rated between 150 and 250 W.

The micro inverters operate at this lower power point, many design issues vital to larger designs are not applicable. Furthermore, the need for the larger ones is generally eliminated. Large electrolytic capacitors can be substituted by more dependable thin-film capacitors in a micro invertor. Cooling loads are avoided as no fans required. Micro inverter connected to a single panel allows it to be detached and tune the output of that panel.

Generally micro inverters have no effect on other panels around it, even if any panel is found to be under-performing. As a whole Micro inverter produces as much as 10 % more power than it would with a string inverter. When shadowing is affected, these improvements can become substantial up to 10 % better output at a minimum, and up to 25% better in some cases.

Furthermore, a single model can be used with a widespread variety of panels, new panels can be added to an array at any time, and do not have to have the same rating of the existing panels.

Micro inverters generate grid-matching power directly and arrays of panels are connected in parallel to each other, and then to the utility grid. This has the advantage of partial production and the inverter cannot take the entire string offline during single panel failure.

The overall array reliability of a micro inverter-based system is significantly greater than a string inverter. Furthermore, when faults happen, they are traceable to a single point, as opposed to an entire string. This not only makes fault separation easier, but exposes minor problems that might not otherwise be visible as single under-performing panel may not affect a long string's output enough to be noticed.

Advantages of Micro Inverters
- Elasticity to partial shading effects as compared to the central and string inverters.
- MPPT at module level
- Highest system plasticity for future expansion
- Minimum DC wiring costs
- Monitoring at module level

Disadvantages of Micro Inverters
- High cost per Watt
- High maintenance costs
- Difficult access for maintenance since the installation is under the PV modules

**String-inverters**

String inverters are connected to multiple solar modules or panels of the PV system. String inverter are the most commonly used inverters in home and commercial solar PV power systems. It is very often fixed on the module mounting structure. Depending on the size of installations, number of strings are connected to the inverter.

In the case of string inverter system each PV modules has its own inverter. A string inverter is an equipment for converting DC to AC power and which is designed for high voltage DC inputs. The rated Maximum DC input power for these inverters will be in the range of 2 – 30kWp.

Advantages of a String Inverter

- Reduced size when compared to central inverters
- Better MPPT capability per string
- Elasticity for future expansion by adding parallel strings
- Short DC wires
- Monitoring at string level

Disadvantages of a String Inverter

- The installation requires special racking for the inverter for each string
- Poor flexibility at partial shading
- Higher cost per Watt than central inverter

## Central Inverters

This is generally used for applications of large PV arrays installed on buildings, industrial facilities as well as ground-mounted solar plants. In this type of inverter, all strings are connected to the DC side and the single AC output is connected to the utility grid. The rated Maximum DC input power for these inverters will be in the range of 50 – 1MWp.

Advantages of a Central Inverter:

The most old-fashioned inverter technology and Easy system design and application. Low cost per Watt and easy approachability for maintenance and troubleshooting

Disadvantages of a Central Inverter

- High DC wiring costs and power loss due to Voltage Drop.
- Single MPPT for the entire PV system
- System output can be severely reduced in case of partial shading and string mismatch
- Difficult to add strings or arrays for future expansion
- Single failure point for the entire system
- Monitoring at array level
- Gigantic size

**Hybrid Inverter**

Conventional inverters store energy systematically in batteries with significant loss. But Hybrid inverters store energy only when there is more production in the PV system than the consumer consumption. This system also allows to choose whether electricity from photovoltaic panels should be stored or consumed through an intelligent control unit.

This is possible through a technique that adds different energy sources. Hybrid inverters operate on both on-grid and off-grid at the same time and backup. A hybrid inverter generally known as a multi-mode inverter. This is an inverter which can simultaneously manage inputs from both solar panels and a battery bank, charging batteries with either solar panels or the electricity grid. The rated Maximum DC input power for these inverters will be in the range from 100 VA onwards.

These inverters are used in off grid mode with the opportunity of linking to a generator. The inverter must be connected to a battery bank and must have true off grid capabilities. But all hybrid inverters are not created equal or can be used in off grid applications.

When used in on-grid or grid-tie with the option of selling energy, there is a need to have the norm compliance of protection and isolation. Use in hybrid mode the inverter functions with a battery bank, but also connected to the grid. This dual functionality is the highlight of hybrid inverters that hence enable energy conservation.

Use in Back-up mode, or storage mode prevents blackouts by switching from on-grid mode to off-grid mode at the moment of electric outage, thereby eliminates network cuts.

**Advantages of Hybrid Inverters**

- All-in-one inverter solution for grid-connected and off grid solar plants
- Frequently intelligent and programmable for maximizing overall system efficiency and savings
- Can be installed without batteries for future expansion

**Disadvantages of Hybrid Inverters**

- Less design flexibility than modular solutions which use separate PV and battery inverters.
- Generally, less efficient than dedicated solar-only or battery-only inverters.

**PV Inverter Selection Criteria**

Some of the important selection criteria for PV inverters are voltage range, efficiency, reliability and protection from climate conditions.

**Voltage range**

The DC input voltage range of the inverter must be well-matched to the voltage of PV System. The inverter datasheet will stipulate the minimum and maximum operating PV array DC voltages.

Number of modules in series should jointly have the DC voltage compactible to inverter DC voltage range. The AC output voltage of the inverter must be identical to the grid voltage.

**Inverter efficiency and reliability**

The main contributor of energy loss in a PV system is inverter. Efficiency is the key performance criteria of PV inverters. This is the quantity of power distributed to the load by the quantity of power taken from the PV array.

The lost energy is not delivered to the load, as the extra power delivered by the PV array to the load will be wasted by the inverter itself. A high efficiency inverter transmits maximum amount of energy to the load or utility grid. The reliability of a PV inverter depends mainly on manufacturing process and technology involved.

**Protection from environmental conditions**

The most significant aspect in selecting an inverter includes factors such as the ambient temperature, chances of dust precipitation and the presence of other contaminants or presence of harmful chemicals. For a definite project these factors must be considered for the inverter selection. It should be established that the environmental stipulations of the inverter have adequate safety margin.

# ANTI ISLANDING PROTECTION

**G**rid-tied inverters are to connect the PV system to the utility grid. The invertor converts DC power to AC power in synchronization with the grid. The main aspects of PV grid interconnection are safety, power quality, and anti-islanding.

Islanding is the condition of powering the network in the event of utility grid failure. Thus, an "island" of powered network within un-powered network is formed even though voltage sources from the electricity utility is disconnected.

This phenomenon can occur due to a faulty condition within the grid or when the grid displays a resonant-load behaviour. In such conditions even if the voltage from the grid network is absent, the resonance between the reactive components (L-C) of the grid will maintain the voltage at the inverter's output terminal. In this condition a current source inverter cannot detect the absence of grid voltage, and if the resistive load matches the power produced by the inverter, the parallel operation is continued forming an "island condition".

This state may prevent automatic re-connection of devices to the system. Furthermore, without strict frequency regulator, the equilibrium between load and generation in the islanded circuit is going to be disturbed, leading to irregular frequencies and voltages.

If an island condition is existing and the utility workers working on the grid may touch live lines when expecting no voltage is present on the line. The probability of electrocution is high in such situations.

After the grid is down, the PV panels still continue to power the line as long as the sunlight is available. This will result in equipment damage and safety risks to technical personnel. This "island" condition in a grid connected solar system continues to produce electricity and feed to the connected network is to be avoided for safety.

Damage to customer equipment could occur if operating limitations differ greatly from anticipated nominal values. In that case, the utility is accountable for the damage. Reclosing the grid onto an active island may affect the utility's equipment, and cause automatic reclosing systems to fail to identify the problem.
Reclosing onto an energized island may cause damage to connected inverters.

For those reasons, distributed generators must sense islanding and directly disconnect from the circuit. This automatic operation is named **anti- islanding**. An island syndrome can be hazardous to safety of line workers and lead to equipment damage or grid collapse.

Most AC grid-ties inverters have anti-islanding feature, so the inverter will reduce power to zero within 2 seconds of the grid shut-down. The requirements for implementation of anti-islanding protection are based on safety standards adopted by utilities. Time interval from detection to grid disconnection indicates, how quickly the inverter is disconnected from the grid.

Detecting an islanding condition demands substantial research. For a given inverter design these differences require competence of different settings of the anti-islanding circuit controls.

In general, the anti-islanding strategies can be classified into passive methods and active methods. Passive methods cover any system that attempts to detect temporary changes on the grid, and use that data for a probabilistic determination of the grid condition, or some other condition that has resulted in a temporary change. The important measurement elements in passive methods are:

**Under/over voltage**

The voltage in an electrical circuit is related to electric current and the applied load. In the case of a grid interruption, the current being supplied by the local source is not likely to match the load so flawlessly as to be able to keep a constant voltage. A system that periodically samples voltage and looks for sudden changes can be used to sense a faulty state.

Under/over voltage finding is normally unimportant to implement in grid-interactive inverters, because the elementary function of the inverter is to match the grid conditions, including voltage. All grid- interactive inverters, by necessity, have the circuitry needed to detect voltage changes. Hence an algorithm is introduced to sense unexpected changes.

Sudden changes in voltage are common incidence on the grid as loads are attached and removed. So, a threshold must be used to evade false interruptions. But non-detection with this method will be alarming, and these systems are generally used along with other detection systems.

**Under/over frequency**

The frequency of the power being carried to the grid is a basic characteristic of the supply, one that the inverters carefully match. When the grid source is lost, the frequency of the power would fall to the natural resonant frequency of the circuits in the island.

Looking for changes in this frequency, like voltage, is easy to implement using existing functionality, and for this reason almost all inverters also look for fault conditions using this method as well. Unlike changes in voltage, it is generally considered highly unlikely that a random circuit would naturally have a natural frequency the same as the grid power. However, many devices deliberately synchronize to the grid frequency, like televisions.

### Rate of change of frequency

In order to decrease the time interval in which an island is detected, rate of change of frequency has been adopted as a detection method. If the rate of change of frequency is greater than a certain value, then solar plant will be automatically disconnected from the grid.

### Voltage phase jump detection

The power factor of normal load is not unity, meaning that they do not receive the voltage from the grid perfectly, but the phase angle deviates slightly. Grid-tie inverters normally have power factors of unity. Changes in phase angle when the grid fails can be used to sense islanding.

Inverters usually track the phase of the grid signal using a phase locked loop (PLL) system. The PLL stays in synchronization with the grid signal by tracking when the signal crosses zero volts. The system will deliver sine-shaped output, varying the current output to the circuit to produce the proper voltage waveform.

As the circuit is providing a current with a smooth voltage output for the known loads, the condition will result in a sudden change in voltage. By the time the waveform is completed and returns to zero, the signal will be out of phase.

The main benefit of this approach is that the shift in phase will occur even if the load exactly matches the supply. On the other hand, the disadvantage is that, common events like starting motors will cause phase hurdles as new impedances are added to the circuit. This compels the system to use fairly large thresholds, reducing its effectiveness.

**Harmonics detection**

Even with noisy sources, like motors, the total harmonic distortion of a grid-connected circuit is generally unmeasurable. The infinite capacity of the grid can filter these distortion events out. Inverters, on the other hand, usually have much higher distortions, as much as 5%.

Thus, when the grid disconnects, the total harmonic distortion of the local circuit will naturally rise to that of the inverters themselves. This provides a very secure method of detecting islanding, because there are usually no other sources of total harmonic changes that would match that of the inverter. Furthermore, components within the inverters themselves, especially the transformers, have non-linear effects that produce exclusive 2nd and 3rd harmonics that are readily measurable.

The drawback of this method is that some loads may filter out the distortion, in the same technique that the inverters employ. If this filtering effect is strong enough, it may reduce the total harmonic distortion below the threshold needed to activate detection.

Modern inverters try to reduce the total harmonic distortion as much as possible, in some cases to unmeasurable limits. Moreover, inverters without a transformer on the "inside" of the disconnect point will make detection more difficult.

### Active methods

Active methods probe the grid by sending signals from the inverter to the grid. It senses a grid failure by injecting test signals into the grid, and then detecting the signal changes.

**Negative-sequence current injection**

Negative-sequence current injection is an active islanding detection method implemented with the help of electronically coupled distributed generation (DG) units. This is based on injecting a negative-sequence current through the voltage-sourced converter controller. The equivalent negative-sequence voltage at the point of common coupling is measured by a unified three- phase signal processor.

The negative-sequence current is injected by a negative-sequence controller. Advanced form of phase-locked loop (PLL) technique is used and which offers a high degree of immunity to noise, and thus permits islanding detection based on injecting a small negative- sequence current.

### Impedance measurement

In this method the overall impedance of the circuit being

supplied by the inverter is measured. This is achieved by slightly "forcing" the current amplitude through the AC cycle, presenting too much current at a given time.

Normally this would have no effect on the measured voltage, as the grid is an effective infinite voltage source. In the event of a disconnection, small forcing would result in a noticeable change in voltage, allowing detection of the island.

**Slip mode frequency shift**

This is one of the modern methods of islanding detection, and also the best. It is based on the principle of shifting the phase of the inverter's output with the grid, with the anticipation that the grid will overpower this signal.

The system depends on the actions of a finely tuned phase-locked loop to become unstable when the grid signal is absent. PLL attempts to adjust the signal back to itself. At the time of grid failure, the system will quickly drift away from the design frequency, ultimately causing the inverter to stop functioning.

The benefit of this method is that it can be applied by using existing invertor circuitry. The main disadvantage is that it requires the inverter always at a low power factor.

**Frequency bias method**

In frequency bias technique a slightly off-frequency signal is injected into the grid. This generates a signal analogous to slip mode, but the power factor remains nearer to that of the grid. The main difficulty is that every inverter would have to approve to shift the signal back to zero at the same point on the cycle.

The Frequency jump type, also known as the "zebra method", inserts signals on an explicit number of cycles in a set pattern. This vividly reduces the chance that external circuits may filter the signal out. This benefit disappears with multiple inverters, unless some other way of synchronizing the patterns is used.

**Utility-based methods**

There are other methods that the utility can detect the conditions and deliberately create conditions in order to make the inverters switch off. A few utility methods available to force the systems offline in the event of a failure is detailed below.

**Manual disconnection**

Small scale solar plants require a mechanical disconnect switch to ensure safe and quick maintenance. In the case of very large plants, a dedicated manually operated hotline with an operator for controlling the shutdown operation is required. In either case, the reaction time is in the order of minutes.

**Automated disconnection**

Physical disconnection could be automated with help of signals sent though secondary means. Power line carrier communications could be used for periodical checking of signals from the grid and disconnecting either on command. This type of system is expensive but highly reliable.

**Transfer-trip method**

Transfer trip is very common safety protection procedure followed by utilities. Utility operators can detect the fault by automated means or close inspection of recloser. This information is transmitted down the line and used for the tripping of properly equipped solar systems. This method requires the grid to be equipped with automated recloser systems, and external communications systems that guarantee the signal will make it through to the recloser.

### Impedance insertion

This is similar to the transfer-trip method, but uses active systems at the head-end of the utility, as opposed to trusting on the network topology. Large capacitor banks are added to the network and allowed to be charged fully. It is usually kept disconnected from live network by a switch. In the occasion of a disaster, the capacitors are switched into the circuit by the utility after a short delay by automatic means.

The capacitors can only supply current for a short period, confirming that the start or end of the pulse they supply will trip the inverters. The capacitor bank has to be large enough to cause changes in voltage that will be sensed. Very large banks of capacitors are expensive and unfavourable to the utility.

## SCADA

Anti-islanding protection can be enhanced with the help of the Supervisory Control and Data Acquisition (SCADA) systems. Alarm can alert, if the SCADA system senses voltage on a line where a disaster is known to be in development. This does not affect the other anti-islanding systems, but may help other systems for quick implementation.

# ROOF TOP SOLAR SYSTEM

The potential of rooftop solar is enormous and this is a very good business opportunity to tap the unutilized solar energy for the commercial and domestic purpose. Rooftop model is having the added advantage of utilising unused roof top in our occupancy for economic activity rather than wasting the costly and useful agricultural land for power generation.

Roof top solar power plant is a green energy source, converts the left-out places of Sun exposed apartment terrace into income generating floor area. Roof top solar systems are of no fuel cost, pollution free and noise free with an added advantage of cooling effect on roof top terrace. It involves less maintenance cost as there are no moving parts. Normal life span of roof top solar panels ranges from 20 years to 25 years.

Roof top solar has become a popular model of energy self-sustainability. This model has double advantage of providing shade for building as well as power with same investment.

Roof top solar is a distributed power generation model

clubbed with small scale income generation for rural household that takes green energy revolution to new heights.

A photovoltaic power plant is carbon negative over its lifespan, as energy generated from the panel counterpoises the need for burning fossil fuels. Even though the Sun doesn't shine always, solar installation gives a reasonably expectable average drop in carbon consumption.

This model can be integrated with grid system for power system stability and reliability. The electricity in this model is generated and consumed at the same premise which makes this solar plant more attractive as distribution loss is incredibly low.

The power- generating capacity of a photovoltaic system is represented in kilowatt peak measured at standard test conditions and a solar irradiation of 1000 W per $m^2$. As per present technological condition, PV modules requires an area between 6-10 $m^2$ per KWp.

The main factors of concern in setting up a roof top solar plant is availability of sunlight. The second factor is roof limitations for system's orientation and tilt in receiving maximum solar radiations. The system's efficiency in converting sunlight to electricity is another factor.

The photovoltaic conversion efficiency is at its maximum if the Sun rays hit the panel vertically. Therefore, PV modules should be oriented to Sun somewhat inclined to south if it is northern hemisphere. The best leaning angle will be subject to the location (including latitude and altitude). The residential solar PV system mostly divided into two classes.

- Off Grid Solar Rooftop System
- Grid connected Rooftop system.

**Off Grid Rooftop System**

A dedicated solar PV system supplying to an exclusive fixed load, not synchronized with the grid, is an Off-grid system. The figure shows the representations of off-grid system

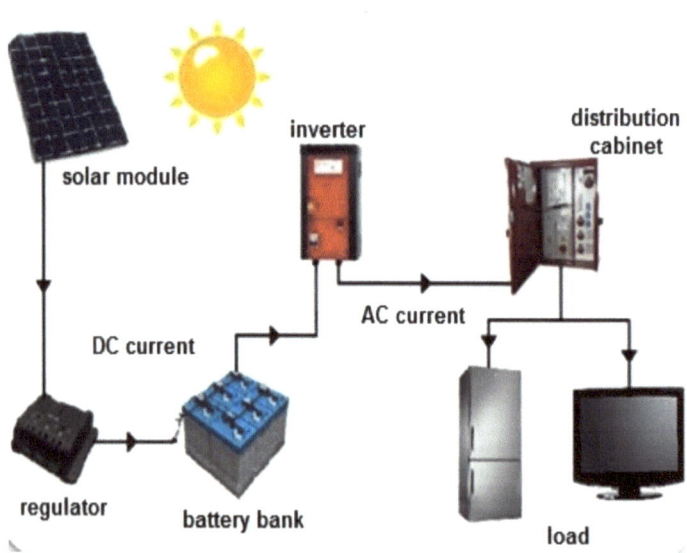

It is an independent system for the generation of electrical energy and is an excellent alternative solution for locations that cannot be connected to the main power grid.

**Grid connected solar PV system**

Grid-connected photovoltaic system is an electricity generating solar PV system of photovoltaic panels, and connected to utility grid. A grid connected PV system fundamentally consists of the PV panels, solar inverters and protections device for automatic shutdown in case of a grid breakdown.

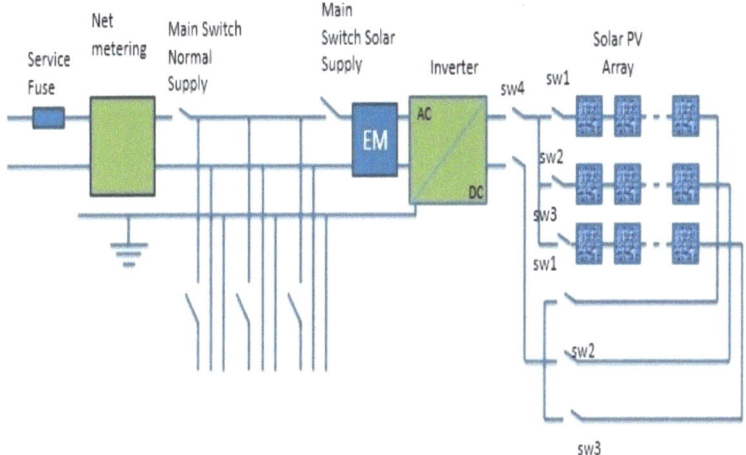

They range from small residential and commercial rooftop systems to large utility-scale solar power stations. Grid-connected PV system supplies the surplus power to the utility grid. The inverter converts the direct current (DC) generated by the modules to alternating current (AC), instantaneously synchronizing the AC output to the AC in the grid. Typical scheme of grid connected solar PV System is shown above.

## Advantages of grid connected System

- A grid-connected photovoltaic power system will reduce the power bill as it is possible to sell extra electricity generated to the local electricity grid.
- Grid-connected PV systems are relatively easier to install as they do not require a battery system.
- Grid interconnection of photovoltaic (PV) power generation systems has the benefit of active utilization of generated power because there are no storage losses involved.

## Community solar systems

Those who are staying in apartment may not have enough open space with sufficient sunlight exposure to install individual solar power system. In such cases the most affordable and cost-effective installation of solar panels in a common area of an apartment will reduce the higher electricity tariff. The electricity usage towards water pumping motors, sewage motors and common area lighting can be done with this common solar system.

The burden of additional electricity bill can be avoided in common maintenance areas with community solar system. By reducing the expenses of common area electricity consumption, the maintenance charge of apartments can be reduced considerably. This will increase the tenant occupancy rate in any apartment and will result in more revenue for the owner.

The resident association of the apartments can take the lead role and a grid integrated system can be proposed for high end consumers of the apartments.

If the individual house owners are not permitted to have separate generating station, then they can go for the option of community solar systems. In such cases the solar panels are installed at a common place and the power generated from these stations are equally divided among individual apartments.

In certain cases, the power generated from the common solar system is fed to the common utility grid and profit of the power generated is divided among the investors. There are different methods for consumers to participate in solar energy production. Few among them are

- Purchasing a solar energy system
- Community or shared solar plants
- Solar power leases
- Solar power purchase agreements with producers

New generation crystalline panels of emerging technologies are used for high power applications in houses and business complex. Homes and businesses complex install rooftop solar system to replace or supplement the grid connected electric supply.

# ROOF TOP SOLAR INSTALLATION

The solar energy that can be harnessed at a particular house is decided by the quantity of solar radiation falling on that area. Photovoltaic electricity can be produced from direct radiation as well as from diffused radiations.

A detailed location study should be conducted before roof top solar plant installation. The first and foremost step in the process of powering residential building with solar energy is auditing the energy consumption of that home. The energy auditing of all home equipment is to be carried out with due consideration to diversity factor and occupancy details.

This will help us to understand where energy efficiency augmentation can be made with the replacement of inefficient appliances.

The house owner should know the exact consumption and usage of electricity in his house.
Once the deficiencies of existing electric equipment are set right, the panel design for appropriate size and capacity can be

started.

The intensity level of radiation decides the solar plants capacity design and panel selection. In order to identify the intensity level of solar radiations a lot of tools and mapping services are available. The solar radiation intensity can be measured with the help of a pyrometer and possible power output from the available Sun exposed area is calculated. With the help of these tools the actual energy that can be tapped is decided and the task of the design part of solar system is initiated.

The production of electricity for the house is depending on quantity of the solar energy reaching over the roof and the type of solar panel that is being installed.

The selection of panel for the residential building should be according to the availability of direct sunlight and diffused sunlight. The type and number of solar panels are selected for suitability and profitability. The arrangements of solar system should be in accordance with maximum Sun exposure.

The shade falling on rooftop may cause hindrance to the harvest of solar energy. Pruning of branches of neighbourhood trees will help to get maximum solar radiation for photovoltaic power generation. In addition to that minute arrangements can be done with civil structural works to get maximum exposure to solar radiation by the panel.

The government declared subsidies and discounts for the

promotion of solar energy projects can be availed for installing solar roof top plants for residential buildings if the plants are installed as per utility regulations. Figure shows the schematic diagram og grid connected solar plant.

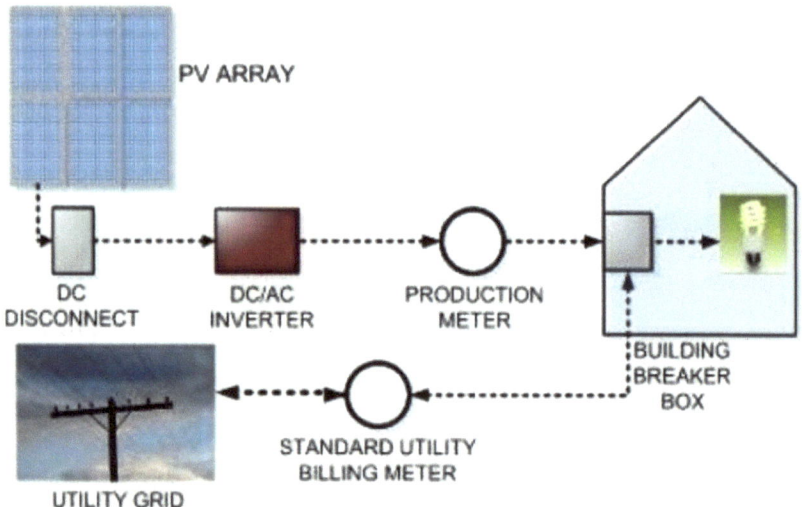

The installation of solar system should be done by an experienced and licenced contractor or installer. If the solar plant is to be connected to electricity grid, then the electric utility owning the grid in the locality should be informed well in advance for layout approval.

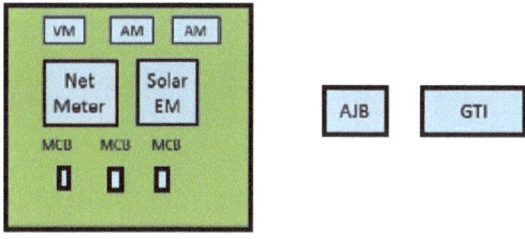

The lay out and the position of isolator, meter and other components should be as per utility norms. Required approval and permission for inter connection may be obtained from the utility in advance. For safety aspects the layout diagram and schematic diagram of solar plant should get approved before starting the work.

Once the work is completed statutory authority should be informed for combined site inspection. The testing parameters include harmonics content level, output voltage levels, frequency range and other system stability parameters. If all the parameters are satisfactory and within the approved limits then the authorities will issue permission for system interconnection through a net meter. Net meter will record the import energy from utility grid and export energy from solar plant for billing purpose.

If the solar systems are installed with the help of a developer, then the developer will help to judge the position of solar panel installation and the plant size required to meet power demand of the house. The developer will ensure the orientation and tilt of solar panel for maximum power generation on daily or seasonal basis. Developer will definitely complete the necessary steps for getting the approval from the utility authorities.

Photovoltaic installations require no maintenance over their entire life is a legend still very persistent in the solar power industry. The reality is quite different and solar power systems like any other technical equipment have abnormalities in their production and installation should be checked and maintained.

Unlike any thermal power station or a factory producing engineering goods, a solar power generating plant is silent in operation as there are no moving components. Except for the humming noise of the transformers, there is no other sound emanating from the plant. However, the plant is potentially dangerous with very high lethal voltages which can cause serious damage to human and equipment of not operated and maintained properly.

Even with latest technology the conversion efficiency of most present-day photovoltaic cells is only about 15 to 20 percent with periodical cleaning of panels. The main objectives of the roof top plant maintenance are to keep the plant running reliably and efficiently as long as possible.

Solar project will provide high return on investment. Return on investment from solar plant is possible between 6 to 7 years with an increased profitability margin in electricity bill. Furthermore, solar project is more environment friendly, that satisfies the commitment to preserve our mother earth.

# AFTERWORD

During last decade photovoltaic technology has shown impressive growth towards technological and economic maturity. Technologists have established an innovative design that could generate more solar energy from minimal area.

The sudden growth in rooftop solar plant installation and its architectural acceptance boosted the demand for solar panels. Cutting edge technological developments have increased the efficiency of solar panels to generate electricity.

Above all, the cost of solar panels has reduced significantly over a period of time and it is the best way to combat global warming. Solar power is environment friendly and easily available green energy source on earth which can end the regime of high-cost fossil fuel.

# ABOUT THE AUTHOR

## Dr Rajan Pullanghad

Dr Rajan Pullanghad, born in 1964, did his graduation in electrical engineering and post-graduation in power system. He did MBA in finance management and PGDBM in energy management.

He took his PhD in 2014 and turned to writing . He has incredible hands-on experience in power sector and renewable energy for the last 30 years in India and abroad. At present he is working as consultant and energy auditor.

He has published many books on renewable energy and power industry management. His works include published books on Solar energy, Energy & Environment, Wind Energy and Utility management.

www.ingramcontent.com/pod-product-compliance
Lightning Source LLC
Chambersburg PA
CBHW040316220526
45473CB00009B/2449